DERMATOLOGY - LABORATORY AND CLINICAL RESEARCH

DERMAL FIBROBLASTS

HISTOLOGICAL PERSPECTIVES, CHARACTERIZATION AND ROLE IN DISEASE

DERMATOLOGY - LABORATORY AND CLINICAL RESEARCH

Additional books in this series can be found on Nova's website under the Series tab.

Additional e-books in this series can be found on Nova's website under the e-book tab.

DERMATOLOGY - LABORATORY AND CLINICAL RESEARCH

DERMAL FIBROBLASTS

HISTOLOGICAL PERSPECTIVES, CHARACTERIZATION AND ROLE IN DISEASE

XIAOWEN BAI
EDITOR

New York

For permission to use material from this book please contact us:
Telephone 631-231-7269; Fax 631-231-8175
Web Site: http://www.novapublishers.com

Library of Congress Cataloging-in-Publication Data

ISBN: 978-1-62618-799-3

Library of Congress Control Number: 2013938559

Published by Nova Science Publishers, Inc. † New York

Contents

Preface

Dermal fibroblasts are the most numerous cells within the dermal layer of skin. The main function of fibroblasts is to preserve the structural integrity of skin through constantly secreting extracellular matrix. Various factors diversely affect the secretion activities of fibroblasts and may lead to varying disease processes (e.g., solar damage and wrinkles). This book gathers important cutting-edge research in the field of fibroblasts covering characterization of fibroblasts and their roles in multiple disease processes such as hypertrophic scarring, aging/photoaging, diabetic wound healing, cancer, and pachydermoperiostosis. In addition, the chapters also deal with the stem cell-like properties of fibroblasts, which sheds new light on the commonly used definition of fibroblasts and provides novel insights into many past studies performed with fibroblast preparations, since the reported effects might be greatly attributable to stem cell content within these fibroblasts. Additionally, this book features an introduction of the reprogramming of dermal fibroblasts into patient-specific induced pluripotent stem cells (iPSCs) that can be converted into varying terminally differentiated cells. Application of iPSCs is a rapidly developing research field because reprogrammed iPSCs from dermal fibroblasts provides a great tool for studying molecular mechanisms of disease, development, and tissue regeneration. This book is certain to have broad appeal and should provide a useful resource for researchers, students, teachers, and physicians interested in the field.

Chapter 1 – Fibroblasts are cells found in the connective tissue and are concerned with tissue maintenance where they sustain structural integrity. They form part of and synthesise the extracellular matrix (ECM) and play a key role in wound healing. Diabetes Mellitus (DM) is a group of metabolic diseases which is characterised by the increase of glucose in the blood. The cause may be due to the lack of production of insulin, the hormone responsible for cellular glucose uptake, or insulin resistance. DM puts patients at risk for a variety of long-term complications, including vascular disease, neuropathy and nephropathy. Together, neuropathy and vascular complications in the lower limbs put patients at jeopardy for developing foot ulcers, a condition which often becomes chronic and necessitates amputation. Low intensity laser irradiation (LILI), otherwise known as phototherapy or photobiostimulation, is a form of light therapy which makes use of low powered lasers or light emitting diodes (LEDs) to stimulate biological processes. Chromophores within the cells absorb the photon energy which leads to a series of chemical reactions which drive cellular metabolism. LILI has expressly been shown to accelerate wound healing, even in conditions such as diabetes. Laser irradiation has been shown to accelerate fibroblast function in a number of stressed conditions; it has been shown to hasten fibroblast migration, increases their proliferation, stimulates the release of nitric oxide (NO) and collagen production, and decreases the release of pro-inflammatory cytokines. This chapter will look at the effect of LILI on fibroblasts, particularly in diabetic wound heaing.

Chapter 2 – The authors examined intracellular calcium ($Ca^{2+}{}_i$) changes and cell proliferation after acute exposure to Angiotensin II in cultured skin fibroblasts from 9 Type 1 diabetic patients with and 9 without nephropathy and from a group of 9 matched normal controls. They found that basal $Ca^{2+}{}_i$ in quiescent fibroblasts was higher in Type 1 diabetic patients with nephropathy than in those without and in controls. Angiotensin II evoked a fast and transient rise in $Ca^{2+}{}_i$, which was higher in fibroblasts from Type 1 diabetic patients with nephropathy (233±17 nmol/L; $p<0.001$) than in those without (159±9) and in normal controls (161±2). Cell proliferation was higher in Type 1 diabetic patients with nephropathy (mean doubling time 3.8±0.8 days; $p<0.05$) than in those without (7.8±2) and normal controls (7.9±1.9). The increase in $Ca^{2+}{}_i$ was inversely related with mean doubling time ($r=-0.45$; $p<0.001$). A significant mitogenic effect of Angiotensin II was found in fibroblasts from Type 1 diabetic patients with nephropathy. Losartan inhibits completely Angiotensin II-induced changes of $Ca^{2+}{}_i$ and proliferation rate in fibroblasts from diabetic patients. In conclusion, this study shows an association between the enhanced responsiveness to Angiotensin II and cell proliferation rate in fibroblasts from Type 1 patients with nephropathy. These abnormalities, persisting in long-term cultured fibroblasts, suggest that the increased susceptibility to nephropathy may reside in an intrinsic determined host-cell response to diabetes.

Chapter 3 – The bulk of human skin is composed of a dense collagen-rich connective tissue, which provides structural and mechanical support of the skin. The collagen-rich connective tissue is produced, organized, and maintained by fibroblasts in the dermis. By creating dermal connective tissue structural and mechanical environment, dermal fibroblasts control skin's homeostasis. The major task of the skin is to protect the human body from harmful environmental factors, such as solar ultraviolet (UV) radiation and micro-organisms. During aging, the dermal connective tissue collagen, the most abundant protein in skin, undergoes progressive alterations, which cause skin connective tissue aging. Age-related alteration of dermal collagen reduces skin dermis structural and mechanical properties and creates a tissue microenvironment that promotes age-related skin diseases, such as poor wound healing and skin cancer. This chapter describes the role of dermal fibroblasts in skin connective tissue aging and in age-related skin diseases.

Chapter 4 – Hypertrophic scar (HTS) following thermal injury is a dermal fibroproliferative disorder (FPD) that leads to considerable morbidity. The development of HTS involves numerous cell types and cytokines with dermal fibroblasts being a key cell. In this work, the properties of dermal fibroblasts in HTS were discussed. First of all, heterogeneous dermal fibroblasts were discussed, particularly focused on dermal fibroblasts from different depth. In normal skin, the properties fibroblasts isolated from different depth using dermatome are different. Cells from the deeper layers were observed to proliferate at a slow rate, but were morphologically larger. The cells from the deeper layers produced more TGF-β1 and CTGF. TGF-β1 producing cells in deeper layers were higher. Cells from the deeper layer tended to transform into myfibroblasts. Fibroblasts from the deeper layers were also found to produce more collagen, but less collagenase. Interestingly, cells from the deeper layers also produced more proteoglycn and versican, but less decorin. Therefore, fibroblasts from the deeper layers of the dermis resemble HTS fibroblasts, suggesting that the deeper layer fibroblasts may play a critical role in the formation of HTS. The immunoinflammatory property of dermal fibroblasts using TLR-4 as an example was also discussed. Furthermore, the phenotype of fibroblasts isolated from HTS was altered as indicated by changed molecular expression such as TGF-β and delayed apoptosis. Understanding heterogeneous property and the role of dermal fibroblasts in inflammatory and immune regulation in post-burn HTS and targeting stromal cell-mediated chronic inflammation offers the potential for novel exciting therapies - not only for dermal scarring, but possibly for other important types of FPD for which effective treatments are seriously lacking.

Chapter 5 – Dermal fibroblasts are essential for the function of normal adult squamous epithelium under physiological conditions and in the course of wound healing. They also play a substantial role during the development of epithelial appendages such as hairs and teeth.

In this chapter, the authors focus on characterization of the fibroblasts located in tumor stroma, with particular stress on the tumors originating from squamous epithelium. These fibroblasts represent an important component of the tumor microenvironment, which is able to influence the biological properties of the tumors including their local aggressiveness and metastatic potential.

They differ from the normal fibroblasts in their transcription profile, as revealed by microarray analyses, and in their crosstalk with epithelial cells. The authors propose therapeutic manipulation of this interaction as a target for a new type of anticancer adjuvant biological therapy.

Chapter 6 – The invasiveness of epithelial cancers arises with age and is known to depend on interactions between the tumor cells and their microenvironment. Precising the exact nature of those interactions deserves further investigations. Human skin is an excellent model for studying genetic and environmental factors contributing to carcinoma development upon aging. The authors' studies on cells isolated from patients presenting high susceptibilty to cancer, such as Xeroderma pigmentosum, or the Nevoid Basal Cell Carcinoma or Gorlin' syndrome, revealed that both epithelial and mesenchymal cells were affected in the absence of genotoxic stress and external stimuli. The authors modelized cancer-like features at the cutaneous level by developing organotypic cultures using the skin cells of patients. Notably, patients' fibroblasts overexpressed factors associated with extracellular matrix modifications (MMP1, MMP3), cell proliferation (KGF/FGF7), or inflammation (SDF1α/CXCL12), all hallmarks of dermal fibroblasts activated in the vicinity of epithelial tumors (Carcinoma-Associated Fibroblasts). Patients' fibroblasts could thus play a key role in early tumor development and aggressiveness. Deciphering the contribution of the stroma to the evolution of epithelial cancer is expected to improve antitumoral treatments.

Chapter 7 – The dermis is a dynamic, supportive connective tissue composed of an extracellular matrix (collagen, elastic fibers, hyaluronic acids, etc.) and ground substance that is mainly formed by dermal fibroblasts. Dermal fibroblasts interact with each other and further with keratinocytes, dendritic cell, monocytes and/or macrophages, neutrophils and lymphocytes, playing a crucial role in regulating skin physiology, such as wound repair and anti-aging.

On the other hand, an aberrant fibroblast phenotype is known to contribute to the pathogenesis of fibrotic disorders, such as systemic sclerosis, localized scleroderma, nephrogenic systemic fibrosis, and keloid scars.

It has recently been revealed that the etiology of a rare genetic disease, pachydermoperiostosis (PDP) or primary hypertrophic osteoarthropathy (PHO), is related to abnormalities in dermal fibroblasts. PDP or PHO is characterized by their unique phenotypes of the skin and bone, such as thick skin and periostitis of long bones. PDP/PHO is attributed to the mutations in the *15-hydroxyprostaglandin dehydrogenase* (*15-PGDH*) gene or the *solute carrier organic anion transporter family, member 2A1* (*SLCO2A1*) gene, which lead to elevated serum prostaglandin E_2. The aim of this chapter is to review the roles of dermal fibroblasts in the pathogenesis of pachydermia in association with Wnt/β-catenin signal pathway and prostaglandin E_2. Enhanced Wnt/β-catenin signaling induced by prostaglandin E_2 seems to contribute to the development of pachydermia by enhanced dermal fibroblast functions.

Chapter 8 – Fibroblasts are plastic adherent mesenchymal cells that play a significant role during tissue development, maintenance, and repair. Meshenchymal stem cells (MSCs) are adherent fibroblast-like cells and are initially isolated from adipose tissue, bone marrow, and skeketal muscle. MSCs are characterized by their extensive proliferative capacity and the ability to differentiate into other cell lineages such as adipocytes and osteoblasts. In contrast to MSCs, fibroblasts are widely considered to be terminally differentiated cells that are primarily responsible for the synthesis and remodeling of extracellular matrix in tissues and do not convert into other types of cells. Although fibroblasts and MSCs have been well studied, differences between these two cell types are not fully understood. Growing experimental evidence demonstrate that skin fibroblasts share similar characteristics with MSCs (e.g., antigen and gene profiles, colony-forming ability, and differentiation potential) as described as follows: (1) fibroblasts and MSCs exhibit a similar spindle-like morphology and express similar antigen profiles: positive for MSC markers (CD44, CD73, and CD105) and fibroblastic markers (collagen I, heat shock protein 47, vimentin, fibroblast surface protein, and α smooth muscle actin), and negative for endothelial cell marker (CD31) and hematopoietic lineage markers (CD14 and CD45); (2) these two types of cells secrete angiogenic growth factors (e.g., VEGF); and (3) both fibroblasts and MSCs show the capacity to form colonies and differentiate into adipocytes, osteoblasts, and chondrocytes *in vitro. In vivo* animal studies also indicate the differentiation potential of fibroblasts. Collectively, these findings suggest that (1) fibroblast preparations from the skin consist of a significant number of stem cells with differentiation potential; (2) skin fibroblasts might play an important role in cutaneous wound healing; and (3) skin fibroblasts can be a promising alternative autologous cell source for tissue regeneration because of their easy access. Importantly, these findings shed new light on the commonly used definition of fibroblasts and provide novel insights into many past studies performed with fibroblast preparations since the reported effects might be greatly attributable to stem cell content within skin fibroblast preparations. This chapter will review the stem cell-like characteristics of dermal fibroblasts.

Chapter 9 – A few hundred human fibroblast strains were established from neonatal foreskins and several strains were selected as good sources for natural type human interferon beta (IFN-β) production in accordance with the authors' criteria such as growth rate, life span, karyological normality, freezing stability, and producing ability. The primary cultures of selected strains had typical fibroblasts morphorology and high IFN-β productivity. The in vitro life span was approximately 60 population doubling level (PDL) and karyotypic analysis indicated that over 90% of the cells had the normal human male diploid karyotype (46, XY). Now, one of the strains was used for producing natural human IFN-β for clinical use. Recently, it was found that the established cell lines could produce a large amount of various cytokines such as IL-6, IL-8, G-CSF, GM-CSF, GRO-α, GRO-β, IL-1α, IL-1β, IL-11, RANTES, and IFN-β in response to the stimulation of double-stranded RNA. The IFN-β and other cytokine productions were enhanced by treating the cells with a low dose of IFN-β. However, the time course of these cytokine productions was different from that of IFN-β. The production peaks of these cytokines appeared after 2 or 3 days of IFN-β production. The cytokine productions were inhibited by treating the cells with anti-human IFN-β monoclonal antibody. These data indicate that (1) various cytokine productions are mediated by previously induced IFN-β production, and (2) human fibroblasts might contribute to the host resistance by producing IFN-β and various physiologically active substances against virus infection, inflammation, or tumor formation.

Chapter 10 – Reprogramming of induced pluripotent stem (iPS) cells has been successfully achieved from mouse and human somatic cells. iPS cells have the capability of indefinite self-renewal and can be differentiated into almost all cell types of the body. This makes them a valuable tool for studying developmental biology, modeling of human genetic diseases, and investigating their therapeutic potential. Specifically, human iPS cells were reprogrammed from dermal fibroblasts from patients with heart diseases, and used to generate patient-specific cardiomyocytes, which have termendous applications in biomedical research, drug development, and clinical practice.

In: Dermal Fibroblasts
Editor: Xiaowen Bai

ISBN: 978-1-62618-799-3

Chapter 1

The Influence of Low Intensity Laser Irradiation on Dermal Fibroblasts and Its Role in Diabetic Wound Healing

*Nicolette N. Houreld**
Laser Research Centre, Faculty of Health Sciences,
University of Johannesburg, South Africa

Abstract

Fibroblasts are cells found in the connective tissue and are concerned with tissue maintenance where they sustain structural integrity. They form part of and synthesise the extracellular matrix (ECM) and play a key role in wound healing. Diabetes Mellitus (DM) is a group of metabolic diseases which is characterised by the increase of glucose in the blood. The cause may be due to the lack of production of insulin, the hormone responsible for cellular glucose uptake, or insulin resistance. DM puts patients at risk for a variety of long-term complications, including vascular disease, neuropathy and nephropathy. Together, neuropathy and vascular complications in the lower limbs put patients at jeopardy for developing foot ulcers, a condition which often becomes chronic and necessitates amputation. Low intensity laser irradiation (LILI), otherwise known as phototherapy or photobiostimulation, is a form of light therapy which makes use of low powered lasers or light emitting diodes (LEDs) to stimulate biological processes. Chromophores within the cells absorb the photon energy which leads to a series of chemical reactions which drive cellular metabolism. LILI has expressly been shown to accelerate wound healing, even in conditions such as diabetes. Laser irradiation has been shown to accelerate fibroblast function in a number of stressed conditions; it has been shown to hasten fibroblast migration, increases their proliferation, stimulates the release of nitric oxide (NO) and collagen production, and decreases the release of pro-inflammatory cytokines. This chapter will look at the effect of LILI on fibroblasts, particularly in diabetic wound heaing.

* Corresponding Author address P.O. Box 17011, Doornfontein, Johannesburg, South Africa, 2028; Email: nhoureld@uj.ac.za_Tel:_+27 11 559-6406 Fax +27 11 559 6884.

Keywords: Diabetes, wound healing, lasers

Introduction

Fibroblasts are specialised cells found in connective tissue where they are responsible for production of the extracellular matrix (ECM) and maintain structural integrity. Morphologically, they appear as large, elongated, spindle-shaped cells with numerous cytoplasmic extensions from the ends of the cell body and abundant rough endoplasmic reticulum. When confluent, they appear as long thin cells aligned in parallel. Fibroblasts play an essential role in wound healing, they migrate to the wound site and ultimately determine how well a wound heals (Porter, 2007). These fibroblasts begin to produce collagen while endothelial cells begin to produce new vasculature. The end result is the production of new tissue, known as scar tissue as its architecture varies somewhat from the original surrounding tissue.

When the process of wound healing is disrupted, or a wound is detained in one of the phases (often the inflammatory phase), a chronic wound often develops. Treatment of these wounds is challenging and requires accurate assessment, prompt treatment, and follow-ups to minimise the potentially debilitating character of these wounds (Sen et al., 2009). Chronic wounds are frequently seen in patients with diabetes mellitus (DM). These chronic wounds often develop as lower limb ulcerations and are due to vascular disease and neuropathy. It is estimated that up to 25% of patients with DM will develop a diabetic foot ulcer (Singh et al., 2005). These patients are also at an increased risk for lower-limb amputations due to infection of the underlying tissue and bone. In the United States approximately 20% of the 3 million hospitalisations each year related to DM is due to diabetic foot ulcers and foot complications (Sen et al., 2009). This places a huge economic burden on society as patients require intensive treatments.

Low-intensity laser irradiation (LILI) refers to the use of low powered lasers to stimulate biological processes. When used as a therapy *in vivo*, such as in pain management or wound healing, it is often referred to as low-level laser therapy (LLLT) or phototherapy. Numerous studies have shown the positive effect of LILI both *in vitro* and *in vivo* and it is used in a wide variety of medical specialities and disciplines from dermatology, dentistry, veterinary science, physiotherapy and somatology, to name but a few. LILI has been shown to stimulate and accelerate the process of wound healing in slow-to-heal wounds, even in cases of chronic diabetic ulcers.

Wound Healing

A wound is a pathological state where the tissue becomes separated or destroyed. Normal wound repair is a series of overlapping, highly co-ordinated events (namely haemostasis, inflammation, proliferation and remodelling) directed at reversing the loss of structural integrity and is controlled by a variety of cells, growth factors, cytokines and metabolic enzymes all released at the wound site (Houreld and Abrahamse 2010a; Houreld and Abrahamse, 2010b). This entire process is directed toward closure of the defect, usually by replacement with scar-forming connective tissue. Following injury, the first phase of wound healing hemostasis is initiated, the coagulation cascade is activated and a clot is formed to prevent localised haemorrhage. Neighbouring tissue releases pro-inflammatory cytokines and growth factors. Inflammatory cells begin to infiltrate into the area and the second phase of inflammation begins. These inflammatory cells release more cytokines attracting more cells to the site. Macrophages in the area undergo phenotypic transition and keratinocytes, fibroblasts and angiogenesis are stimulated (Meszaros et al., 2000; Mosser and Edwards, 2008) and the third stage, proliferation, is initiated. During this stage, the ECM is formed. Fibroblasts and endothelial cells within the reparative dermis support capillary growth and collagen and granulation tissue formation (Guo and DiPietro, 2010). The final stage, remodelling, is finally entered into, where the wound undergoes contraction and remodelling of the ECM to that of normal tissue. Wound contraction begins approximately 8-10 days post-injury. Fibroblasts apply tension to the surrounding ECM and form cell-to-cell links further contributing to wound contraction. The quality of wound healing is dependent on the equilibrium between collagen formation and degradation. Sequential deposition of the different types of collagen peaks at about three weeks and the wound tensitle strength is about 60% of the previously unwounded skin (Falanga, 2005).

Collagen is one of the most abundant proteins found *in vivo* and is a major component of the ECM where it offers structural support. During wound healing, collagen synthesis is controlled and maintained by a fine balance between matrix synthesis and degradation. This is regulated by matrix degrading enzymes, matrix metalloproteases (MMPs, proteolytic enzymes responsible for collagen degradation), and their inhibitors, tissue inhibitor metalloproteinases, (TIMPs). Normal wound healing processes require a certain level of these enzymes, however if there is an imbalance there may be an increase in collagen degradation and a decrease in construction, resulting in chronic wounds. MMPs known to be important in normal wound healing include the collagenases (MMP-1, -8, -13) and gelatinases (MMP-2, -9), (Armstrong and Jude, 2002).

The integrated events of normal wound healing must occur in the correct sequence at the correct time and continue for the correct duration (Guo and DiPietro, 2010). Delayed wound healing arises when these events are disrupted.

A chronic wound is one which has failed to proceed through the normal wound healing process and often do not heal within three months. Acute and chronic wounds commonly enter a state of pathologic inflammation and the majority of chronic ulcers are associated with ischemia, diabetes, venous stasis disease, or pressure (Guo and DiPietro, 2010).

Diabetic Wound Healing

DM is a group of metabolic diseases which is characterised by an increase in blood glucose levels. The cause of these increased levels may be due to the lack of production of insulin, the hormone responsible for cellular glucose uptake, or insulin resistance. The World Diabetes Foundation estimated that 285 million people, or 6.4% of the world's population, had diabetes in 2010 and that this figure would increase to 438 million, or 7.8%, by the year 2030. Global prevalence in 2010 was at 6.4%, however prevalence varied from 10.2% in the Western Pacific to 3.8% on the African continent, however Africa is expected to have the highest incidence (World Diabetes Foundation). This is mainly due to the change in lifestyle as more African countries become Westernised.

Diabetes is one of the leading causes of premature death and the number of deaths in 2010 showed a 5.5% increase from the year 2007 (World Diabetes Foundation). DM puts patients at risk for a variety of long-term complications, including vascular disease, neuropathy and nephropathy. Together, neuropathy and vascular complications in the lower limbs put patients at jeopardy for developing foot ulcers, a condition which often becomes chronic and necessitates amputation. Around 50% of all non-traumatic amputations are as a result of DM (Schindl et al., 1998). Due to sensory loss, foot pathologies often go unnoticed in diabetic patients and these frequently develop into deeper ulcers which become slow-to-heal. Further complications such as infection and gangrene arise, which often necessitate amputation due to spread to underlying tissue and bone. It is estimated that 15-25% of patients will develop foot ulcers (Pendensy 2010; Singh et al., 2005; Richard et al., 2011), of which 6% of these will require hospitalisation to treat these ulcers (Carrington et al., 2001). Around 15-20% of these patients will require lower limb amputation (Pendensy, 2010). To further highlight the seriousness of diabetes associated lower-limb amputations, the 5 year mortality rate following amputation stands at 39-80% (Reiber, 2001). DM and its associated complications impact heavily on the patient, their family, health care departments and governments. In the USA, approximately $25 billion is spent annually on the treatment of diabetic complications (Fernyhough et al., 2010). The treatment of chronic wounds is becoming more of a burden due to the increase in health care costs, an aging population and an increase in the incidence of diabetes (Sen et al., 2009).

Chronic wounds are rarely seen in otherwise healthy individuals and are commonly seen in patients with diabetes (Sen et al., 2009). Numerous studies have shown that wound healing in DM does not follow the normal process and sequence. Increased levels of glucose in the blood of diabetic patients lead to the formation of chronic wounds (Figure 1). There is impaired wound healing and angiogenesis and decreased wound strength (Francis-Goforth et al., 2010), and due to vascular disease, there is decreased blood flow which further complicates the process. There is a disruption in clot formation and all phases in the inflammatory phase are dysregulated (Peppa et al., 2009; Chen et al., 2005), often with a prolonged and excessive inflammatory response, which is further amplified by hypoxia (Guo and DiPietro, 2010).

Formation of the ECM is a crucial step in wound healing and provides structural integrity to tissue. In diabetic wound healing, there is an irregularity in the formation of the ECM due to the disruption of ECM-growth factor interactions and abnormal fibroblasts which display poor migration, proliferation, adhesion and growth factor production (Francis-Goforth et al., 2010; Amos et al., 2010). There is also an imbalance between MMPs and TIMPs. The loss of collagen, which is associated with diabetes, can be due to decreased levels of its synthesis, enhanced metabolism or a combination of both (Arul et al., 2007). The non-healing diabetic foot wound displays elevated MMP activity, and there is a 30 to 60-fold increase and is predominantly MMP-2 and -9 (Guo and DiPietro, 2010; Trengove et al., 1999).

Nitric oxide (NO) is significantly reduced in chronic ulcers and impaired healing of diabetic wounds is thought to be related to this decrease (Schäffer et al., 1997; Bulgrin et al., 1995; Schwentker et al., 2002). Burrow et al., (2007) demonstrated that normal skin fibroblasts produce more NO than diabetic human skin fibroblasts, and that there was a direct relationship between NO levels and MMP expression; there was an inverse proportion such that low NO concentrations resulted in high MMP levels. Their experiments also showed that diabetic human skin fibroblasts expressed higher levels of MMP-8 and -9. Weller and Finnen (2006) showed that the application of a topical acidified nitrite cream, which delivered NO to the wound, augmented the healing process in diabetic mice. In contrast, Nakai et al., (2003) found an increase in NO in HaCat cells (immortalised human keratinocyes) which were grown in hyperglycemic conditions (25 mmol/l).

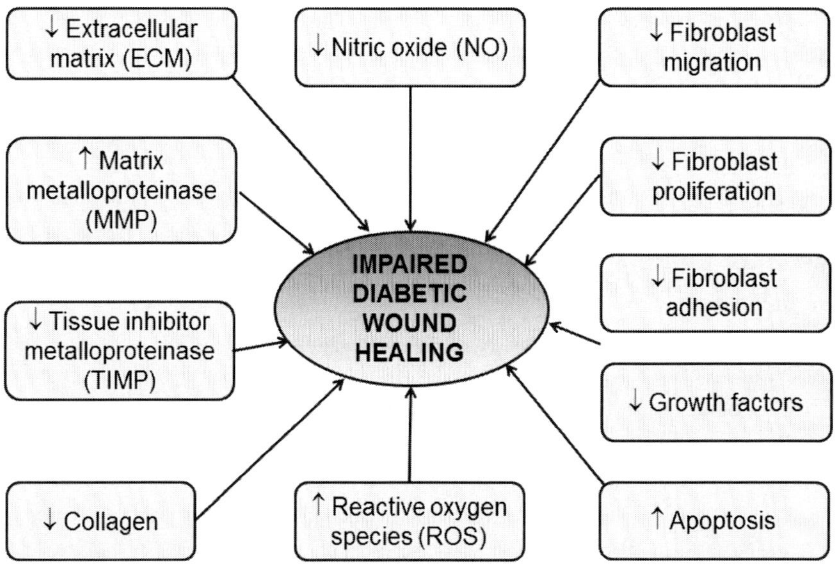

Figure 1. Some of the underlying pathogenesis of impaired diabetic wound healing.

Treatment of Diabetic Wounds

The treatment of chronic wounds is a challenging one. Chronic wounds lead to patient disability, which worsen the outcomes of wound healing, and so the cycle goes on and on. Chronic wounds require accurate assessment, prompt treatment and suitable follow-ups to minimise the potential long-term disability (Sen et al., 2009). Another aspect of pain further complicates matters. Patients experience persisting pain as well as pain during changes of the wound dressings. This results in lower activity, depression, irritation and reduced social activity (Sen et al., 2009). Management of the diabetic foot is multidisciplinary and can become problematic. Treatment is both local (treating of the diabetic foot) and systemic (glycemic control). Ulcer management is extensive and can encompasses mechanical and surgical debridement, management of the wound base, antibiotic therapy to treat infection, revasularisation, prophylactic foot surgery, mechanical off loading, accommodative orthotics and an alteration of footwear (Singh et al., 2005; Richard et al., 2011).

Alternative healing treatments under investigation include hyperbaric oxygen therapy (HBOT), ultrasound, growth factors, negative pressure and stem cell therapy. HBOT entails delivering 100% oxygen at pressures above one atmosphere and increases the amount of oxygen dissolved in the blood, and has been used to treat a variety of wounds (Eskes et al., 2011; O'Reilly et al., 2011). HBOT has been used as an adjunctive treatment for diabetic foot ulcers; however its evidence of efficiency is limited (O'Reilly et al., 2011). Kaya and colleagues (2009) treated 184 patients with diabetic foot ulcers with HBOT. Sixty three percent of patients responded to treatment, while 17% showed no improvement and 21% underwent amputations. Complications associated with HBOT are not common but can include claustrophobia, ear, sinus or lung damage, temporary worsening of short sightedness and oxygen poisoning (O'Reilly et al., 2011).

Therapeutic ultrasound involves the use of acoustic pulsed energy and has been reported to promote protein synthesis, calcium uptake, cellular proliferation, collagen deposition and DNA synthesis, and increase wound tensile strength (Altomare et al., 2009). Although positive effects have been found, its clinical use in wound healing is still being investigated due to conflicting reports. Altomare et al., (2009) found that ultrasound accelerated healing by increasing wound contraction and re-epithelialisation and better organisation of the ECM. It was also found to have an anti-inflammatory response. However, it has not been found to be effective when ischemic wounds were treated, in fact there was a decrease in wound contraction, re-epithelialisation, myofibroblastic differentiation and angiogenesis and increased inflammation.

The use of growth factors applied to ulcers has been found to be somewhat successful. Living skin equivalents (e.g. Apligraf and Dermagraph) have been found to be effective, however they require weekly applications, have a short shelf life and are expensive. Becaplerim (Regranex®, platelet-derived growth factor-B homodimer, PDGF-BB, Systagenix Wound Management, Gargrave, UK) is approved for use in the treatment of diabetic ulcers, however there is only a 15% increase in healing as compared to the placebo group (Mulder et al., 2009). Mulder et al., (2009) showed that the topical application of GAM501 (adenoviral vector containing the human PDGF-BB gene incorporated into a biocompatible matrix of bovine collagen) resulted in a rapid reduction in wound size and wound closure in the majority of patients. Negative pressure wound therapy (NPWT) makes use of controlled negative pressure using a vacuum-assisted closure device (VAC) to remove fluid from open wounds through a sealed dressing and tubing which is connected to a collection container. NPWT has been found to reduce infection and wound discharge, and increase vascularity, cytokine expression, wound healing, tissue granulation and epithelialisation (Nain et al., 2011; Zelen et al., 2011). The use of stem cells in wound repair is being used as a means to reinitiate the hosts tissue's wound healing cascade. Adipose derived stromal cells (ASCs) have been used for such experiments and are seeded onto acellular dermal scaffolds and have been found to secrete various growth factors and improve wound healing both *in vivo* and *in vitro* (Amos et al., 2010). ASCs have also been found to stimulate fibroblast proliferation, migration and epithelialisation and augment the production of matrix factors (Amos et al., 2010).

Phototherapy

Lasers, an acronym for Light Amplification by Stimulated Emmision of Radiation, produce light in the non-ionising part of the electromagnetic spectrum. Laser light is monochromatic, meaning it is light of a single wavelength; coherent, meaning that it is organised and all the photons are in phase and; directional, meaning that the beam of light is tight, strong and concentrated. Energy contained within light is expressed in joules (J) and its fluence or energy density in joules per square centimetre (J/cm^2). Power is the rate at which work is performed and is measured in watts (W). The power density or irradiance (i.e. light concentration) is the amount of optical energy incident upon a surface, divided by the area of the surface being irradiated and is measured in watts per centimetre square (W/cm^2). Wavelength is the distance from crest to crest of an electromagnetic wave and is measured in nanometres (nm).

A number of clinical applications have been found for lasers in a variety of medical specialities, such as, but not limited to, dentistry, dermatology, osteology, physiotherapy, acupuncture, surgery, photodynamic cancer therapy (PDT), chiropractic and veterinary science. Phototherapy, or LILI, involves the use of laser light therapy to stimulate cellular processes. LILI has been shown in a number of studies to stimulate wound healing both *in vivo* and *in vitro*, and exerts it's effect on a number of cell types, such as fibroblasts (Houreld and Abrahamse, 2008; Burd et al., 2005; Chen et al., 2011) keratinocytes (Gavish et al., 2004; Burd et al., 2005), endothelial cells (Chen et al., 2008), neural cells (Giuliani et al., 2009; Oron et al., 2007), chondrocytes (Baumann et al., 2006), osteoblasts (Stein et al., 2005), osteoclasts (Aihara et al., 2006), monocytes (Gulsoy et al., 2006; Hemvani et al., 2005) and smooth muscle cells (Gavish et al., 2006), to name but a few. Phototherapy treats in a non-destructive and non-thermal manner (Karu et al., 2008).

The effect of LILI on tissue is dependent not only on the laser characteristics, but also on the properties of the target tissue. The depth at which laser light penetrates the target tissue is dependent on its wavelength. Visible laser light interacts preferentially with pigmented tissue; while blue-green light is absorbed by pigments making its penetration depth low (Ohshiro, 1991), visible red light is less absorbed by blood and has a higher penetration depth. Laser light from the neodymium yttrium-aluminum-garnet (Nd:YAG) laser is generally absorbed by proteins, thus its penetration rate in tissue is quite high. Light from a diode laser, particularly the gallium-aluminum-arsenide laser (GaAIAs, 830 nm) has the highest penetration rate (Ohshiro, 1991). Thus the choice of laser is dependent on the penetration depth required. Wavelengths in the 600 to 700 nm range are chosen for treating superficial tissue, while wavelengths between 780 and 950 nm are chosen for deeper-seated tissues (Hamblin and Demidova, 2006).

LILI has been found to promote healing in both soft and hard tissue. Irradiation at 980 nm has been shown to stimulate wound healing and the formation of new bone in diabetic rats (Park and Kang, 2012). In a study conducted by Peplow and colleagues (2012) it was shown that irradiation of diabetic mice with a 660 nm laser did not have a hypoglycemic effect and that the effect of wound healing is due to cellular and biochemical changes in the immediate wound environment. Laser irradiation increases the healing of splintered wounds in diabetic mice by increasing reepithelisation and granulation tissue formation (Chung et al., 2010), as well as repairing tenotomised achilles tendon (Nouruzian et al., 2013). Al-Watban (2009) treated Sprague-Dawley rats and concluded that LILI at 633 nm administered three times a week at a fluence of 4.71 J/cm^2 should be used for diabetic burn wounds, and a fluence of 2.35 J/cm^2 be used for diabetic ulcers. Irradiation of diabetic ulcers that failed to respond to other forms of treatments with combined wavelengths (660 and 890 nm, 3 J/cm^2) has been shown to heal faster than placebo irradiated ulcers (Caetano et al., 2009; Minatel et al., 2009). LILI has even been shown to inhibit bacterial growth. In a study by Enwemeka and colleagues (2009) they showed a decrease in methicillin-resistant *Staphylococcus aureus*

(MRSA) treated with blue laser light (470 nm). It has been suggested that infected wounds should be treated with 408 nm to clear the infection, followed by 730 nm to speed up healing of the wound (Ankri et al., 2010).

Effect of Low Intensity Laser Irradiation on Fibroblasts

LILI has been used in numerous studies on fibroblasts to determine the effects of LILI seen in wound healing and more specifically diabetic wound healing (Figure 2). Diabetic wounds show decreased fibroblast migration and proliferation. LILI has been shown to increase both these processes. When wounded dibatic induced fibroblast cells are irradiated at 632.8 nm with 5 J/cm^2, there was increased migration of cells, with complete wound closure as compared to control cells which showed incomplete wound closure (Houreld and Abrahamse, 2008).

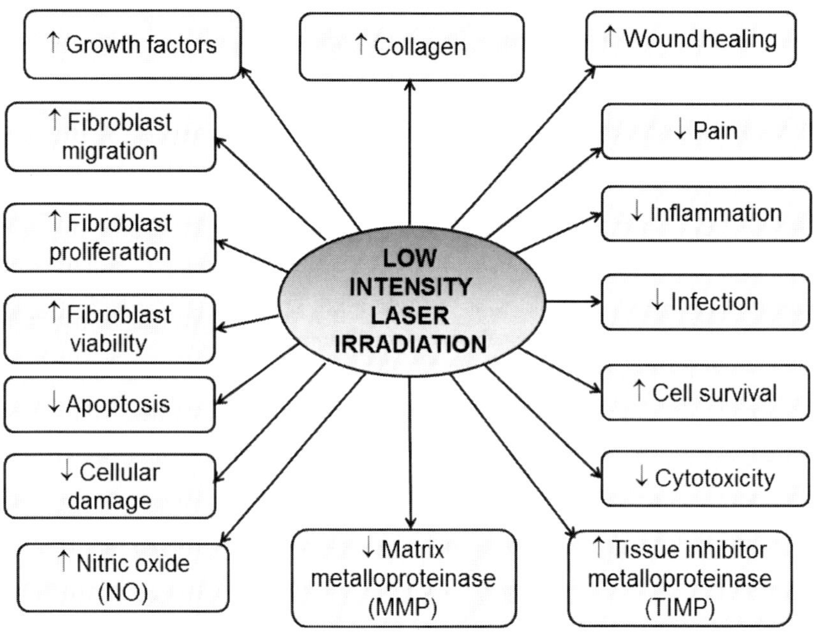

Figure 2. Some of the effects of low intensity laser irradiation.

The same cells irradiated at 830 nm also showed increased migration, however the rate of migration was slower than that seen in cells irradiated with 632.8 nm. LILI was also shown to increase the migration and proliferation of fibroblast cells in the presence of hydroxyurea, a potent anti-proliferative agent (Zungu et al., 2009). Laser irradiation has also been shown to increase the proliferation of hypoxic fibroblast cells (Sekhejane et al., 2011). The effect of LILI on the proliferation of fibroblast cells appears to be dependant on a variety of laser paramaters such as wavelength, power density, fluence, number of irradiations and duration between irradiations. Hakki and Bozkurt (2012) found that LILI of human gingival fibroblasts using a 904 nm diode laser with various laser settings (2 W at 20 J/cm^2; 1.5 W at 15 J/cm^2; and 0.3 W at 6 J/cm^2) did not effect the proliferation of these cells over an 80 hour period. Almeida-Lopez and colleagues (2001) showed that irradiation at the same fluence (2 J/cm^2) with different wavelengths and power densities had different effects. Irradiation with infrared laser light (50 mW) significantly increased stressed fibroblast proliferation as compared to irradiation using visible red laser light (10 mW). However, when the same power output for the lasers was used, no differences in proliferation between infrared and visible red lasers were noted. On the other hand, when Azevedo et al., (2006) irradiated fibroblasts with a 660 nm diode laser (2 J/cm^2) at different power densities, they noted that cells irradiated at a density of 10 mW (142.85 mW/cm^2) had significantly higher proliferation than the same cells irradiated with a power density of 29 mW (428.57 mW/cm^2). Pereira et al., (2002) irradiated fibroblasts to a 102 mW diode laser with a wavelength of 904 nm at a fluence of 3 to 5 J/cm^2. A fluence of 3 and 4 J/cm^2 significantly increased fibroblast proliferation, while 5 J/cm^2 had no effect. Other studies have shown that irradiation with visible red light at a fluence of 2.5, 3 or 5 J/cm^2 significantly increased fibroblast proliferation (Frigo et al., 2010; Taniguchi et al., 2009; Houreld and Abrahamse 2007a), while irradiation at higher fluences (16 J/cm^2) inhibited fibroblast viability and proliferation (Houreld and Abrahamse 2007a; Houreld and Abrahamse 2007b). When fibroblast cells were irradiated on the same day, or two consecutive days, more damage was inflicted, with a decrease in proliferation and is due to a cumulative effect (Houreld and Abrahamse 2007c). It was also noted that the time of measuring biological responses post-irradiation plays a role (Hawkins and Abrahamse 2007a; Hawkins and Abrahamse 2007b).

Growth factors released by fibroblasts are important in wound healing, and are involved in signaling. Diabetic wounds have shown a decrease in the production of growth factors, which hinders the healing process (Amos et al., 2010). Basic fibroblast growth factor (bFGF) is a mitogen and chemoattractant and accelerates the formation of granulation tissue and re-epithelialisation. Damante et al., (2009) showed that LILI at 780 nm (3 and 5 J/cm^2) significantly increased bFGF, while irradiation at 660 nm showed no significant increase. bFGF is thought to be stored intracellularly or bound to heparin sulfate in the ECM. The reason for this increase in bFGF might be that infrared laser light affects cell membranes and induce molecular vibrations and rotation, thus causing the the release of bFGF. The differences seen between these two wavelengths might be explained by differences in data collection and experimental procedures, as well as laser parameters. Even irradiation of fibroblast cells to red light emitting diodes (LED, 625–635 nm) increased bFGF secretion (Huang et al., 2007).

Insulin-like growth factor 1 (IGF1), also known as somatomedin C, plays an important role in growth and can stimulate the formation of collagen and is also a chemotactic to fibroblasts. Vascular endothelial growth factor (VEGF) stimulates vasculogenesis and angiogenesis, while transforming growth factor-beta (TGF-β) can promote collagen synthesis. Saygun et al., (2008) showed that LILI of fibroblasts at 685 nm (2 J/cm^2) stimulated the release of bFGF, IGF1 and the receptor of IGF-1 (IGFBP3). Hakki and Bozkurt (2012) found that irradiation of human gingival fibroblasts using an infrared 904 nm diode laser with various laser settings (2 W at 20 J/cm^2; 1.5 W at 15 J/cm^2; and 0.3 W at 6 J/cm^2) increased mRNA expression of IGF, VEGF, and TGF-β, and postulated that this increase could be one of the contributing factors to increased healing seen in LILI. Heat-shock proteins (HSP) are proteins involved in the folding and unfolding of other proteins, they act as intra-cellular molecular chaperones. HSPs are named according to their molecular weight and are expressed when cells are under stress. Following thermal stress HSPs enhance the ability of cells to deal with the accumulation of abnormally folded proteins and participates in the synthesis of new proteins (Helbig et al., 2009). Interleukin (IL)-6, a pro- and anti-inflammatory cytokine, has been linked to the pathogenesis of diabetes (Cambell et al., 1991; Morohoshi et al., 1995), however it also linked to cellular proliferation and differentiation. Dang et al., (2010) irradiated human skin fibroblast cells with a 532 (green visible light) or 1064 nm laser (near-infrared light) at a dose of 1.5 J/cm^2 and found that irradiation at 1064 nm upregulated TGF-β, while irradiation at 532 nm upregualted Hsp70 and IL-6. It has been shown that irradiation of diabetic wounded fibroblast cells (632.8 nm, 5 J/cm^2) increased IL-6 and bFGF expression, which resulted in an increase in migration and proliferation of diabetic induced fibroblast cells (Houreld and Abrahamse 2007a; Houreld and Abrahamse, 2010a).

Deposition of the ECM is an important step in the wound healing process. There is a carefully controlled balance between collagen degradation and deposition. Clinical and experimental evidence has shown that diabetic wounds do not follow an orderly and reliable progression of wound healing (Falanga, 2005). Diabetic wounds have shown an imbalance between MMPs and TIMPs and a reduaction in collagen deposition. Hyperglycemia alters the the balance of MMP concentrations and proteolytic activity (Falanga, 2005). Lasers such as the Nd:YAG laser has been effectively used for the treatment of skin wrinkles. Dang et al., (2010) irradiated human skin fibroblast cells with a 532 (green visible light) or 1064 nm laser (near-infrared light) at a dose of 1.5 J/cm^2 and found that there was an upregualtion of type I and III procollagen, TIMP1 and TIMP2, and decreased expression of MMP1 and MMP2. Thus, irradiation at both wavelengths accelerated collagen synthesis and inhibited collagen degradation. The authors hypothesised that the upregulation of TGF-β, Hsp70 and IL-6, might be partly responsible for the increased collagen synthesis seen in these cells. On the other hand, Orringer et al., (2005) and Oh et al., (2007) showed that laser irradiation at 585 nm increased MMP levels, while irradiation at 1320 nm had no effect on MMP levels. Hakki and Bozkurt (2012) saw an increase in collagen type I when fibroblast cells were irradiated at 904 nm (0.3 W) with 6 J/cm^2. In contrast, Pereira et al., (2002) and Marques et al., (2004) showed that irradiation at 3 J/cm^2 with a wavelength of 904 nm had no effect on procollagen. It was shown that irradiation using these parameters resulted in ultrastructural changes to the mitochondria and endoplasmic reticulum, suggesting a disturbance in protein synthesis and delivery (Pereira et al., 2002). In a study on smooth muscle cells, LILI (780 nm) was shown to stimulate collagen synthesis and modulate equilibrium between these matrix remodeling enzymes (Gavish et al., 2006). Irradiation of diabetic wounded fibroblast cells *in vitro* at 660 nm with 5 J/cm^2 produced a significant increase in collagen type I (Ayuk et al., 2012). Irradiation of fibroblast cells to red LEDs increased collagen type I (Huang et al., 2007). McDaniel et al., (2010) conducted gene expression studies on ECM and adhesion molecules in fibroblast cells exposed to varying ratios of 590/870 nm. They showed that different expression profiles were obtained for each of the ratios. The greatest increase in collagen type I and decrease in collagenase was noticed when a ratio of 75/25% (590/870 nm) was used.

Another factor which contributes to impaired diabetic wound healing is the increase in reactive oxygen species (ROS) and pro-inflammatory cytokines. There is an overproduction of the cytokines tumor necrosis factor alpha (TNF-α) and IL-1β (Brandner et al., 2008). TNF-α has also been shown to be an accelerator of diabetes (Hotamisligil et al., 1993), and is associated with apoptosis. IL-1β and TNF-α has been shown to stimulate MMPs and decrease

TIMPs (Brandner et al., 2008). Hyperglycemia induces oxidative stress in diabetic cells, leading to programmed cell death. ROS and NO are also important mediators in wound healing, and their effects are dependant on their concentrations. NO has been found to be decreased in diabetic wounds (Burrow et al., 2007). Nuclear factor kappa B (NF-kB) is an important transcription factor and regulates the expression of a number of genes involved in inflammation and cell survival (Chen et al., 2011). Irradiation of hypoxic or diabetic wounded fibroblast cells at 636 nm with 5 J/cm^2 resulted in a decrease in apoptosis, IL-1β and TNF-α and translocation of NF-kB to the nucleus (Sekhejane et al., 2011). Frigo et al., (2010) also showed that irradiation of fibroblast cells at 660 nm (2.5 W/cm^2, 3 J/cm^2) reduced apoptosis. Irradiation of diabetic wounded fibroblast cells at 830 nm (5 J/cm^2) produced an immediate increase in NO and ROS, and a decrease in apoptosis, IL-1β and TNF-α (Houreld and Abrahamse 2010b). Chen and colleagues (2011) found that irradiation of murine embryonic fibroblasts (810 nm) activated NF-kB via the generation of ROS, resulting in a decrease in apoptosis and pushing cells into survival mode. Lim et al., (2011) showed that irradiation at 635 nm produced a decrease in ROS and an increase in angiogenesis in conditions of hypoxia/ischemia. Diabetic fibroblast cells in culture show increased DNA damage and cytotoxicity as compared to normal fibroblast cells. However when irradiated, these increases were no longer evident (Houreld and Abrahasme 2007c Houreld and Abrahamse 2007d; Mbene et al., 2009), thus LILI is able to reverse the damage inflicted in diabetic cells.

Conclusion

Wound healing is the outcome of a number of biological processes all working in synergy to restore the loss of structural integrity. When this well-orchestrated process is disrupted, chronic wounds can develop. The treatment of such wounds is challenging, and there is a great burden on the patient and caregivers. Despite a wide variety of products and therapies available to treat diabetic ulcers, there is no single treatment or cure sufficiently robust for long-term prevention (Reiber and Raugi, 2005). Many wounds fail to heal despite adequate implementation of the various wound care techniques. There is a need for additional research into alternative therapies for wound healing in diabetes. Early diagnosis, constant monitoring and appropriate treatment are essential.

Therapies based on photochemical conversion are not laser-specific and conventional light sources generating the applicable wavelength can be used (Karu, 1999). Laser light is used due to its properties of monochromaticy, coherence, wavelength tenability, simplicity of use and ease of delivery through fibre-optic cables (Karu, 1999). Although the use of phototherapy is not an established treatment modality, it is gaining a wide amount of interest and has been proven to be useful. A vast amount of research on the use of phototherapy *in vivo* and its underlying effects and mechanisms at a cellular level are being conducted. New research and current advances in wound care encompassing laser irradiation have led to an improvement in the treatment of diabetic foot ulcers and improved the quality of patients' lives. Phototherapy has shown to shorten the period of wound healing where all other therapies have been unsuccessful, as well as alleviating pain and inflammation. With no reported side-effects, this therapy is well worth investigating and more randomised well controlled clinical studies, and studies on the underlying meachanisms are needed.

By studing the effects of LILI *in vitro*, we have been trying to understand the underlying mechanisms involved in LILI assisted wound healing. These studies have shown that LILI has an effect on various cell types, including fibroblasts. LILI has been shown to stimulate a number of processes involved in wound healing in fibroblast cells. There is an increase in migration, cellular proliferation, NO, collagen deposition and growth factors and a decrease in MMPs, inflammatory cytokines, cytotoxicity and DNA damage. *In vivo*, there has been an increase in wound closure and wound tensile strength, increased angiogenesis, re-epithelisation and pain attenuation. LILI has even been shown to eradicate micro-organisms. Not all studies have shown that LILI stimulates fibroblasts. Some researchers found no effect on fibroblast proliferation, collagen synthesis and cellular migration (Hakki and Bozkurt, 2012; Webb and Dyson, 2003; Colver and Riestley, 1989; Pogrel et al., 1997). The effect of LILI on these cells is dose and laser dependent, as well as the physiological state of the cells at the time of irradiation. New research and current advances in wound care encompassing laser irradiation have led to an improvement in the treatment of diabetic foot ulcers and improved the quality of patients' lives.

References

Aihara, N., Yamaguchi, M., and Kasai, K. (2006). Low-energy irradiation stimulates formation of osteoclast-like cells via RANK expression in vitro. *Lasers Med. Sci. 21,* 24-33.

Almeida-Lopes, L., Rigau, J., Zangaro, R.A., Guidugli-Neto, J., and Jaeger, M.M. (2001). Comparison of the low level laser therapy effects on cultured human gingival fibroblasts proliferation using different irradiance and same fluence. *Lasers Surg. Med. 29,* 179–184.

Altomare, M., Nascimento, A.P., Roma-Souza, B. Amadeu, T.P., and Monte-Alto-Costa, A. (2009). Ultrasound accelerates healing of normal wounds but not of ischemic ones. *Wound Rep. Reg. 17,*825-831.

Al-Watban, F.A.H. (2009). Laser Therapy Converts Diabetic Wound Healing to Normal Healing. *Photomed. Laser Surg.27,* 127–135.

Amos, P.J., Kapur, S.K., Stapor, P.C., Shang, H., Bekiranov, S., Khurgel, M., Rodeheaver, G.T., Peirce, S.M., and Katz, A.J. (2010). Human adipose-derived stromal cells accelerate diabetic wound healing: impact of cell formation and delivery. *Tissue Engineering: Part A 16,* 1595-1606.

Armstrong, D.G., and Jude, E.B. (2002). The role of matrix metalloprotrinases in wound healing. *J. Am. Podiatr. Med. Assoc. 92,* 12-18.

Ankri, R., Lubart, R., and Taitelbaum, H. (2010). Estimation of the Optimal Wavelengths for Laser-Induced Wound Healing. *Laser Surg. Med. 42,* 760–764.

Arul, V., Kartha, R., and Jayakumar, R.A. (2007). therapeutic approach for diabetic wound healing using biotinylated GHK incorporated collagen matrices. *Life Sciences 80*, 275–284.

Ayuk, S.M., Houreld, N.N., and Abrahamse, H. (2012). Collagen froduction in wounded fibroblastsin response to low intensity laser irradiation (LILI) at 660 nm. *Diabetes Tech. Ther.* 14, 1110-1117.

Azevedo, L.H., de Paula Eduardo, F., Moreira, M.S., de Paula Eduardo, C., and Marques, M.M. (2006). Influence of different power densities of LILT on cultured human fibroblast growth: a pilot study. *Lasers Med. Sci. 21,* 86–89.

Baumanna. M., Jorgensen, B., Rohde, E., Bindig, U., Muller, G., and Scheller, E.E. (2006). Influence of wavelength, power density and exposure time of laser radiation on chondrocyte cultures – An in-vitro investigation. *Medical Laser Application 21,* 191-198.

Brandner, J.M., Zacheja, S., Houdek, P., Moll, I., and Lobman, R. (2008). Expression of matrix metalloproteinases, cytokines, and connexins in diabetic and non-diabetic human keratinocytes before and after transplantation into an ex vivo wound healing model. *Diabetes Care 31,* 114–120.

Bulgrin J.P., Shanbani M., Chakravarthy D., and Smith D.J. (1995). Nitric oxide synthesis is suppressed in steroid-impaired and diabetic wounds. *Wounds 7,* 48.

Burd, A., Zhu, N., and Poon, V.K.M. (2005). A study of Q-switched Nd:YAG laser irradiation and paracrine function in human skin cells. *Photodermatol. Photoimmunol. Photomed. 21,* 1–7.

Burrow, J.W., Koch, J.A., Chuang, H., Zhong, W., Dean, D.D., and Sylvia, V.L. (2007). Nitric oxide donors selectively reduce the expression of matrix metalloproteinases-8 and -9 by human diabetic skin fibroblasts. *J. Surg. Res. 140,* 90-98.

Caetano, K.S., Frade, M.A.C., Minatel, D.G., Santana, L.A., and Enwemeka, C.S. (2009). Phototherapy Improves Healing of Chronic Venous Ulcers. *Photomed. Laser Surg.. 27,* 111–118.

Campbell, I.L., Kay, T.W., Oxbrow, L. and Harrison, L.C. (1991) Essential role for interferon-gamma and interleukin-6 in autoimmune insulin-dependent diabetes in NOD/Wehi mice. *J. Clin. Invest. 87,* 739-742.

Carrington, A.L., Abbott, C.A., Griffiths, J., Jackson, N., Johnson, S.R., Kulkarni, J., Van Ross, E.R., and Boulton, A.J. (2001). A foot care program for diabetic unilateral lower limb amputees. *Diabetes Care 24,* 216–221.

Chen, C-H., Arany, P.R., Huang, Y.Y., Tomkinson, E.M., Sharma, S.K., Kharkwal, G.B., Saleem, T., Mooney, D., Yull, F.E., Blackwell, T.S., and Hamblin M.R. (2011). Low-level laser therapy activates NF-kB via generation of reactive oxygen species in mouse embryonic fibroblasts. *PLOS/One 6,* e22453.

Chen, C-H., Hung, H-S., and Hsu S-H. (2008). Low-energy laser irradiation increases endothelial cell proliferation, migration, and eNOS gene expression possibly Via PI3K signal pathway. *Lasers Surg. Med. 40,* 46–54.

Chen, J., Kasper, M., Heck, T., Nakagawa, K., Humpert, P.M., Bai, L., Wu, G., Zhang, Y., Luther, T., Andrassy, M., Schiekofer, S., Hamann, A., Morcos, M., Chen, B., Stern, D.M., Naworth, P.P. and Bierhaus, A. (2005) Tissue factor as a link between wounding and tissue repair. *Diabetes 54,* 2143-2154.

Chung, T-Y,. Peplow, P.V., and Baxter G.D. (2010). Laser photobiomodulation of wound healing in diabetic and non-diabetic mice: effects in splintered and unsplintered wounds. *Photomed. Laser Surg. 28,* 251-261.

Colver, G.B., and Riestley, G.C. (1989). Failure of a helium–neon laser to affect components of wound healing in vitro. *Br. J. Dermatol. 121,* 179–186.

Damante, C.A., De Micheli, G. Miyagi, S.P., Feist, I.S., and Marques, M.M. (2009). Effect of laser phototherapy on the release of fibroblast growth factors by human gingival fibroblasts. *Lasers Med. Sci. 24,* 885-891.

Dang, Y., Ye, X., Wneg, Y., Tong, Z., and Ren, Q. (2010). Effects of 532-nm and 1,064-nm Q-switched Nd:YAG lasers on collagen turnover of cultured human skin fibroblasts: a comparative study. *Lasers Med. Sci. 25,* 719-726.

Enwemeka, C.S., Williams, D., Enwemeka, S.K., Hollosi, S. and Yens, D. (2009) Blue 470-nm Light Kills Methicillin-Resistant Staphylococcus aureus (MRSA) in Vitro. *Photomed. Laser Surg. 27,* 221–226.

Eskes, A.M., Ubbink, D.T., Lubbers, M.J., and Lucas, C. (2011). Hyperbaric oxygen therapy: solution for difficult to heal acute wounds? Systematic review. *World J. Surg. 35*, 535-542.

Falanga, V. (2005). Wound healing and its impairment in the diabetic foot. *The Lancet 366,* 1736-1743.

Fernyhough, P., Chowdhury, S.K.R., and Schmidt, R.E. (2010). Mitochondrial stress and the pathogenesis of diabetic neuropathy. *Expert Rev. Endocrinol. Metab. 5*, 39-49.

Francis-Goforth, K.N., Harken, A.H., and Saba, J.D. (2010). Normalization of diabetic wound healing. *Surgery 147*, 446-449.

Frigo, L., Favero, G.M., Campos Lima, H.J., Augusto Maria, D., Bjordal, J.M., Joensen, J., Iversen, V.V., Labat Marcos, R., Antonio Parizzoto, N., and Lopes-Martins R.A. (2010). Low-level laser irradiation increases fibroblast cell proliferation and reduces cell death in a dose-dependent manner. *Photomed. Laser Surg. 28*, S151-S156.

Gavish, L., Asher, Y., Becker, Y., and Kleinman, Y. (2004). Low level laser irradiation stimulates mitochondrial membrane potential and disperses subnuclear promyelocytic leukemia protein. *Lasers Surg. Med. 35,* 369-376.

Gavish, L., Perez, L., and Gertz, S.D. (2006). Low-Level Laser Irradiation Modulates Matrix Metalloproteinase Activity and Gene Expression in Porcine Aortic Smooth Muscle Cells. *Lasers Surg. Med. 38*, 779–786.

Giuliani, A., Lorenzini, L., Gallamini, M., Massella, A., Giardino, L., and Calzà, L. (2009). Low infra red laser light irradiation on cultured neural cells: effects on mitochondria and cell viability after oxidative stress. *BMC Complementary and Alternative Medicine 9,* 8.

Gulsoy, M., Ozer, G.H., Bozkulak, O., Tabakoglu, H.O., Aktas, E., Deniz, G., and Ertan, C. (2006). The biological effects of 632.8-nm low energy He–Ne laser on peripheral blood mononuclear cells in vitro. *J. Photochem. Photobiol. B: Biology 82,* 199-202.

Guo, S., and DiPietro, L.A. (2010). Factors affecting wound healing. *J. Dent. Res 89*, 219-229.

Hakki, S.S., and Bozkurt, S.B. (2012). Effects of different setting of diode laser on the mRNA expression of growth factors and type I collagen of human gingival fibroblasts. *Lasers Med. Sci. 27,* 235-331.

Hamblin, M.R., and Demidova, T.N. (2006). Mechanisms of low level laser therapy. In: Hamblin, M.R., Waynant, R.W., Anders, A., editors. Proceedings of The International Society for Optical Engineering; 2006 Jan 22-24; San Jose, California, USA: pp. 614001-1–614001-12.

Hawkins, H., and Abrahamse, H. (2007a) How long after laser irradiation should cellular responses be measured to determine the laser effect? *J. Laser Appl. 19,* 74-83.

Hawkins, H., and Abrahamse, H. (2007b). Time-dependent responses of wounded human skin fibroblasts following phototherapy. *J. Photochem. Photobiol. B: Biology 88,* 147–155.

Helbig, D., Bodendorfm M.O., Grunewald, S., Kendler, M., Simon, J.C., and Paasch, U. (2009). Immunohistochemical investigation of wound healing in response to fractional photothermolysis. *Journal of Biomedical Optics 14,* 064044-1 - 064044-8.

Hemvani, N., Chitnis, D.S., and Bhagwanani, N.S. (2005). Helium-Neon and nitrogen laser irradiation accelerates the phagocytic activity of human monocytes. *Photomed. Laser Surg. 23,* 571-574.

Hotamisligil, G.S., Shargill, N.S., and Spiegelman, B.M. (1993). Adipose expression of tumor necrosis factor-alpha: direct role in obesity linked insulin resistance. *Science 259,* 87–91.

Houreld, N., and Abrahamse, H. (2007a). Irradiation with a 632.8 nm Helium-Neon laser with 5 J/cm2 stimulates proliferation and expression of interleukin-6 in diabetic wounded fibroblast cells. *Diabetes Technol. Ther. 9,* 451-459.

Houreld, N., and Abrahamse, H. (2007b). In vitro exposure of wounded diabetic fibroblast cells to a Helium-Neon laser at 5 and 16 J/cm^2. *Photomed. Laser Surg. 25,* 78-84.

Houreld, N., and Abrahamse, H. (2007c). Effectiveness of Helium-Neon Laser Irradiation on Viability and Cytotoxicity of Diabetic-Wounded Fibroblast Cells. *Photomed. Laser Surg. 25,* 474-481.

Houreld, N., and Abrahamse, H. (2007d). Cellular damage in diabetic wounded fibroblast cells following phototherapy at 632.8, 830 and 1064 nm. *Laser Chemistry 2007* Article ID 80536, 9 pages doi:10.1155/2007/80536.

Houreld, N., and Abrahamse, H. (2008). Laser light influences cellular viability and proliferation in diabetic-wounded fibroblast cells in a dose- and wavelength-dependent manner. *Lasers Med. Sci. 23,* 11-18.

Houreld, N., and Abrahamse, H. (2010a). Low intensity laser irradiation stimulates wound healing in diabetic wounded fibroblast cells (WS1). *Diabetes Technol. Ther. 12(12),* 971-978.

Houreld, N., and Abrahamse, H. (2010b). Irradiation at 830nm stimulates nitric oxide production and inhibits pro-inflammatory cytokines in diabetic wounded fibroblast cells. *Lasers Surg. Med. 42,* 494-502.

Huang, P-J., Huang, Y-C., Su, M-F., Yang, T-Y., Huang, J-R., and Jiang, C-P. (2007). *In vitro* observations on the influence of copper peptide aids for the LED photoirradiation of fibroblast collagen synthesis. *Photomed. Laser Surg. 25,* 183-190.

Karu, T. (1999). Primary and secondary mechanism of action of visible to near-IR radiation on cells. *J. Photochem. Photobiol. B: Biology 49,* 1-17.

Karu, T.I., Pyatibrat, L.V., Kolyakov, S.F., and Afanasyeva, N.I. (2008). Absorption measurements of cell monolayers relevant to mechanisms of laser phototherapy: reduction or oxidation of cytochrome c oxidase under laser irradiation at 632.8 nm. *Photomed. Laser Surg. 26,* 593-599.

Kaya, A., Aydin, F., Altay, T., Karapinar, L., Ozturk, H., and Karakuzu, C. (2009). Can major amputation rates be decreased in diabetic foot ulcers with hyperbaric oxygen therapy? *Int. Orthop. 33,* 441-446.

Lim, W.B., Kim, J.S., Ko, Y.J., Kwon, H., Kim, S.W., Min, H.K., Kim, O., Choi, H.R., and Kim, O.J. (2011). Effects of 635nm Light-Emitting Diode Irradiation on Angiogenesis in CoCl2-Exposed HUVECs. *Laser Surg. Med. 43,* 344–352.

Marques, M.M., Pereira, A.N., Fujihara, N.A., Nogueira, F.N., and Eduardo, C.P. (2004). Effect of low-power laser irradiation on protein synthesis and ultrastructure of human gingival fibroblasts. *Lasers Surg. Med. 34,* 260–265.

Mbene, A.B., Houreld, N.N., and Abrahasme, H. (2009). DNA damage after phototherapy in wounded fibroblast cells irradiated with 16 J/cm^2. *J. Photochem. Photobiol. B: Biology 94,* 131-137.

McDaniel, D.H., Weiss, R.A., Geronemus, R.G., Mazur, C., Wilson, S., and Weiss, M.A. (2010). Varying Ratios of Wavelengths in Dual Wavelength LED Photomodulation Alters Gene Expression Profiles in Human Skin Fibroblasts. *Laser Surg. Med. 42,* 540–545.

Meszaros, A.J., Reichner, J.S., and Albina, J.E. (2000). Macrophage-induced neutrophil apoptosis. *J. Immunol. 165,* 435-441.

Minatel, D.G., Frade, M.A.C., França, S.C., and Enwemeka, C.S. (2009). Phototherapy Promotes Healing of Chronic Diabetic Leg Ulcers That Failed to Respond to Other Therapies. *Laser Surg. Med. 41,* 433–441.

Morohoshi, M., Fujisawa, K., Uchimura, I. and Nurnano, F. (1995) The effect of glucose and advanced glycosolation end products on IL-6 production by human monocytes. *Ann. NY Acad. Sci. 17,* 562-570.

Mosser, D.M., and Edwards, J.P. (2008). Exploring the full spectrum of macrophage activation. *Nat. Rev. Immunol. 8,* 958-969.

Mulder, G., Tallis A.J., Marshall, V.T., Mozingo, D., Phillips, L., Pierce, G.F., Chandler, L.A., and Sosnowski, B.K. (2009). Treatment of nonhealing diabetic foot ulcers with a platelet-derived growth factor gene-activated matrix (GAM501): results of a phase ½ trial. *Wound Rep. Reg. 17,* 772-779.

Nain, P.S., Uppal, S.K., Garg, R., Bajaj, K., and Garg, S. (2011). Role of negative pressure wound therapy in healing of diabetic foot ulcers. *J. Surg. Tech. Case Rep. 3,* 17-22.

Nakai, K., Fujii, S., Yamamoto, A., Igarashi, J., Kubota, Y., and Kosaka, H. (2003). Effects of high glucose on NO synthesis in humam keratinocyte cell line (HaCaT). *Journal of Dermatological Science 31,* 211-218.

Nouruzian, M., Alidoust, M., Bayat, M., Bayat, M., and Akbari, M. (2013). Effect of low-level laser therapy on healing of tenotomized achilles tendon in streptozotocin-induced diabetic rats. *Lasers Med. Sci.* 28, 399-405.

Oh, J., Kim, N., Seo, S., and Kim, I.H. (2007). Alteration of extracellular matrix modulators after nonablative laser therapy in skin rejuvenation. *Br. J. Dermatol. 157,* 306–310.

Ohshiro, T. (1991). An introduction to LLLT. In: Ohshiro, T., Calderhead, R.G., editors. Progress in Laser Therapy. Selected Papers from the October 1990 ILTA Congress; 1990 October; John Wiley and Sons, West Sussex, England pp. 36–47.

O'Reilly, D., Linden, R., Fedorko, L., Tarride, J.E., Jones, W.G., Bowen, J.M., and Goeree, R. (2011). A prospective, double-blind, randomized, controlled clinical trial comparing standard wound care with adjunctive hyperbaric oxygen therapy (HBOT) to standard wound care only for the treatment of chronic, non-healing ulcers of the lower limb in patients with diabetes mellitus: a study protocol. *Trials 12,* 69.

Oron, U., Ilic, S., De Taboada, L., and Streeter, J. (2007). Ga-As (808 nm) laser irradiation enhances ATP production in human neuronal cells in culture. *Photomed. Laser Surg. 25,* 180-182.

Orringer, J.S., Voorhees, J.J., Hamilton, T., Hammerberg, C., Kang, S., Johnson, T.M., Karimipour, D.J., and Fisher, G. (2005). Dermal matrix remodeling after nonablative laser therapy. *J. Am. Acad. Dermatol. 53,* 775–782.

Park, J.J., and Kang, K.L. (2012). Effect of 908-nm GaAIAs diode laser irradiation on healing of extraction sockets in streptozotocin-induced diabetic rats; a pilot study. *Lasers Med. Sci. 27,* 223-230.

Pendsey, S.P. (2010). Understanding diabetic foot. *Int. J. Diabetes Dev. Ctries 30,* 75-79.

Peppa, M., Stavroulakis, P., and Raptis, S.A. (2009). Advanced glycoxidation products and impaired diabetic wound healing. *Wound Repair Regen 17,* 461-472.

Peplow, P.V., Chung, T.Y., and Baxter G.D. (2012). Laser photostimulation (660 nm) of wound healing in diabetic mice is not brought about by ameliorating diabetes. *Lasers Surg. Med. 44,* 26-29.

Pereira, A.N., Eduardo, C.P., Matson, E., and Marques, M.M. (2002). Effect of low-power laser irradiation on cell growth and procollagen synthesis of cultured fibroblasts. *Lasers Surg. Med. 31,* 263–267.

Pogrel, M.A., Chen, J.W., and Zhang, K. (1997). Effects of low-energy gallium–aluminum–arsenide laser irradiation on cultured fibroblasts and keratinocytes. *Lasers Surg. Med. 20,* 426–432.

Porter, S. (2007). The role of the fibroblast in wound contraction and healing. *Wounds 3,* 33-40.

Reiber, G.E. (2001). Epidemiology of foot ulcers and amputations in the diabetic foot. In: Bowker JH, Pfeifer MA, editors. The Diabetic Foot. St Louis, Mo: Mosby, pp 13-32.

Reiber, G.E., and Raugi, G.J. (2005). Preventing foot ulcers and amputations in diabetes. The Lancet *366,* 1676-1677.

Richard, J.L., Sotto, A., and Lavigne, J.P. (2011). New insights in diabetic foot infection. *Wound Journal of Diabetes 2,* 24-32.

Saygun, I., Karacay, S., Serdar, M., and Ural, A.U. (2008). Effects of laser irradiation on the release of basic fibroblast growth factor (bFGF), insulin like growth factor-1 (IGF-1), and receptor of IGF-1 (IGFBP3) from gingival fibroblasts. *Lasers Med. Sci. 23,* 211-215.

Schäffer, M.R. Tantry, U., van Wesep, R.A., and Barbul, A. (1997). Nitric oxide metabolism in wounds. *Journal of Surgical Research 71,* 25-31.

Schwentker, A., Vodovotz, Y., Weller, R., and Billiar, T.R. (2002). Nitric oxide and wound repair: role of cytokines? *Nitric Oxide 7,* 1-10.

Schindl, A., Schindl, M., Schon, H., Knobler, R., Havelec, L., and Schindl, L. (1998). Low-intensity laser irradiation improves skin circulation in patients with diabetic microangiopathy. *Diabetes Care 21,* 580 – 584.

Sekhejane, P.R., Houreld, N.N., and Abrahamse, H. (2011). Irradiation at 636 nm positively affects diabetic wounded and hypoxic cells *in vitro. Photomed. Laser Surg. 29,* 521-530.

Sen, C.K., Gordillo, G.M., Roy, S., Kirsner, R., Lambert, L., Hunt, T.K., Gottrup, F., Gurtner, G.C., and Longaker, M.T. (2009). Human skin wounds: A major and snowballing threat to public health and the economy. *Wound Repair Regen 17,* 763-771.

Singh, N., Armstrong, D.G., and Lipsky, B.A. (2005). Preventing foot ulcers in patients with diabetes. *JAMA 293,* 217-228.

Stein, A., Benayahu, D., Maltz, L. and Oron, U. (2005) Low-Level Laser Irradiation Promotes Proliferation and Differentiation of Human Osteoblasts *in Vitro. Photomed. Laser Surg. 23,* 161-166.

Taniguchi, D., Dai, P., Hojo, T., Yamaoka, Y., Kubo, T., and Takamatsu, T. (2009). Low-energy laser irradiation promotes synovial fibroblast proliferation by modulating p15 subcellular localization. *Lasers Surg. Med. 41,* 232-239.

Trengove, N.J., Stacey, M.C., Macauley, S., Bennett, N., Gibson, J., Burslem, F., Murphy, G., and Schultz, G. (1999). Analysis of the acute and chronic wound environments: the role of proteases and their inhibitors. *Wound Repair Regen 7,* 442-452.

Webb, C., and Dyson, M. (2003). The effect of 880 nm low level laser energy on human fibroblast cell numbers: a ppossible role in hypertrophic wound healing. *J. Photochem. Photobiol. B: Biology 70,* 39-44.

Weller, R., and Finnen, M.J. (2006). The effects of topical treatment with acidified nitrite on wound healing in normal and diabetic mice. *Nitric Oxide 15,* 395-399.

World Diabetes Foundation http://www.worlddiabetesfoundation.org/ composite-35.htm accessed on 16 July 2012

Zelen, C.M., Stover, B., Nielson, D., and Cunningham, M. (2011). A prospective study of negative pressure wound therapy with integrated irrigation for the treatment of diabetic foot ulcers. *Eplasty 11,* e5.

Zungu, I.L., Mbene, A.B., Hawkins Evans, D., Houreld N.N., and Abrahamse, H. (2009). Phototherapy promotes cell migration in the presence of hydroxyurea. *Lasers Med. Sci. 24,* 144-150.

In: Dermal Fibroblasts
Editor: Xiaowen Bai

ISBN: 978-1-62618-799-3

Chapter 2

Abnormal Calcium Release by Angiotensin II and Growth Rate in Skin Fibroblasts from Type I Diabetic Patients with Diabetic Nephropathy

Renato Millioni[1], Giulio Ceolotto[1], Roberto Trevisan[2] and Elisabetta Iori[1*]

[1]Department of Medicine, Via Giustiniani, University of Padova, Padova, Italy

[2]Diabetes Unit, A.O. Ospedali Riuniti, Largo Barozzi, Bergamo, Italy

Abstract

We examined intracellular calcium ($Ca^{2+}{}_i$) changes and cell proliferation after acute exposure to Angiotensin II in cultured skin fibroblasts from 9 Type 1 diabetic patients with and 9 without nephropathy and from a group of 9 matched normal controls. We found that basal $Ca^{2+}{}_i$ in quiescent fibroblasts was higher in Type 1 diabetic patients with nephropathy than in those without and in controls. Angiotensin II evoked a fast and transient rise in $Ca^{2+}{}_i$, which was higher in fibroblasts from Type 1 diabetic patients with nephropathy (233±17 nmol/L; $p<0.001$) than in those without (159±9) and in normal controls (161±2). Cell proliferation was higher in Type 1 diabetic patients with nephropathy (mean doubling time 3.8±0.8 days; $p<0.05$) than in those without (7.8±2) and normal controls (7.9±1.9). The increase in $Ca^{2+}{}_i$ was inversely related with mean doubling time ($r=-0.45$; $p<0.001$). A significant mitogenic effect of Angiotensin II was found in fibroblasts from Type 1 diabetic patients with nephropathy. Losartan inhibits completely Angiotensin II-induced changes of $Ca^{2+}{}_i$ and proliferation rate in fibroblasts from diabetic patients. In conclusion, this study shows an association between the enhanced responsiveness to Angiotensin II and cell proliferation rate in fibroblasts from Type 1 patients with nephropathy. These abnormalities, persisting in long-term cultured

* Corresponding Author address: Department of Medicine, University Hospital of Padova, Italy; Email: elisabetta.iori@unipd.it.

fibroblasts, suggest that the increased susceptibility to nephropathy may reside in an intrinsic determined host-cell response to diabetes.

Keywords: Type 1 diabetes mellitus, diabetic nephropathy, skin fibroblasts, angiotensin II

Introduction

Diabetic nephropathy (DN) develops in about 25% of type 1 diabetes mellitus (T1DM) patients over 20 years of standard glycaemic control, and is the strongest predictor of cardiovascular disease and premature death in these patients (Nathan et al., 2009). Effective interventions in predisposed individuals should be initiated very early, and will hopefully result in the delay of progression or perhaps in the prevention of DN. Microalbuminuria has been considered to reflect an early stage of DN and is currently the best available non-invasive predictor of DN risk. However some investigations have cast doubt on this by demonstrating that microalbuminuria often regresses to normal and that some T1DM patients with persistent microalbuminuria already exhibit advanced renal structural damages (Ismail et al., 1999; Caramori et al., 2000). Despite growing knowledge about DN pathogenesis, we are still lacking of valuable biomarkers for the early prediction of this disease onset and progression. Diabetic nephropathy has been defined by glomerular lesions including glomerular basement membrane thickening, mesangial expansion and hyalinosis of afferent and efferent arterioles (Steffes et al., 1992; Najafian et al., 2009; Toyoda et al., 2007). Accumulation of extracellular matrix (ECM) is the central abnormality in DN (Steffes et al., 1992). Renal fibroblasts are the main producers of the ECM and their secretion of collagen type IV increases upon stimulation with the transforming growth factor beta (TGF-β1) or high glucose (Lam et al., 2004). Studies directed at the understanding of the early cellular mechanisms responsible for these changes in different renal cell types of predisposed subjects should be very informative. On the other hand, this approach is limited by the availability of kidney biopsies. To circumvent this problem, other cell types have been considered as alternative model. In this regards, human fibroblasts, easily obtained through a skin biopsy, have been largely employed as an in vitro model to study pathophysiological mechanisms of diabetes mellitus, also associated with the development of DN. They offer a useful model for the investigation of intrinsic (possibly genetic) defects of cell function, since fibroblasts are actively involved in the process of renal and cardiovascular fibrosis and in the development of target organ damage in hypertensive and diabetic patients. Even if we cannot be sure that the same abnormalities found in skin fibroblasts would be present in kidney cells as well, many typical cellular features associated with DN have been already observed in cultured fibroblasts. These studies show that cell abnormalities in addition to the metabolic abnormalities associated with hyperglycaemia may play a role in the pathogenesis of DN. Cultured skin fibroblasts from Type 1 diabetic patients with DN exhibit distinctive phenotype differences compared with cells from patients with normoalbuminuria and normal control subjects, such as an increased DNA synthesis (Trevisan et al., 1992; Lurbe et al., 1996), an increased activity of Na^+/H^+ exchange (Trevisan et al., 1992; Lurbe et al., 1996), greater PKC activity (Iori et al., 2003) and a lower Ca-pump-mediates Ca^{2+} efflux (Podestà et al., 2001). Using a proteomic approach, we recently showed that several proteins were differentially expressed between cultured skin fibroblasts from T1DM patients with and without established

DN (Puricelli et al., 2006; Millioni et al., 2010; Iori et al., 2008; Millioni et al., 2008; Tessari et al., 2007), evidencing an association between this diabetic complication and the constitutive expression of some proteins involved in glucose metabolism, protein turnover and cytoskeleton assembly. Angiotensin II (Ang II) plays a key role in mediating the haemodynamic and structural manifestations of diabetic nephropathy (Cooper, 2001) but the precise mechanism of this effect remains unclear. Experimental data support the hypothesis that Ang II has also non-haemodynamic effects on matrix metabolism that may contribute to matrix accumulation in DN (Leehey et al., 2000; Singh et al., 2003). In rat mesangial cells, it was shown that Ang II production was stimulated by hyperglycaemia in association to TGF-β1 production and mediated the inhibitory effects of high glucose on matrix degradation (Singh et al., 2003). In addition, Ang II has been shown to decrease matrix degradation by inhibiting the activity of matrix metalloproteinases-2 (MMP-2) in rat mesangial cells; this effect was mediated via the AT1 receptor because it could be blocked by losartan (Singh et al., 1999). Since Ang II exerts its effect through the mobilization of intracellular calcium (Ca^{2+}_i), any abnormality in this ion handling may be relevant in the pathogenesis of cardiovascular and renal complication of type 1 diabetic patients. We tested the hypothesis that an alteration in Ca^{2+}_i mobilisation after exposure to Ang II was associated to an increase of proliferation rate in cells from type 1 diabetic patients with nephropathy.

Materials and Methods

Subjects

Details of cohort selection have been reported previously (Iori et al., 2003). For this study we selected 9 Type 1 (insulin-dependent) diabetic patients with overt diabetic nephropathy, defined as an urinary albumin excretion (AER) persistently greater than 200 μg/min. Nine normal subjects and nine long-term Type 1 diabetic patients without family history of hypertension and of diabetes with normal albumin excretion rate (AER < 20 μg/min) served as control subjects (Table I). All subjects were of Caucasian origin and gave their informed consent to the study, which was approved by the Committee on Ethical Practice of Padua's Hospital. Mean blood pressure was calculated as diastolic blood pressure plus one third pulse pressure. Glomerular filtration rate (GFR) was determined by plasma clearance of ^{51}Cr-EDTA in the diabetic patients only. Three timed overnight urine samples were collected for measurement of urinary AER and the median value was used for classification.

Cell Culture

Skin biopsies were taken by excision under local anaesthesia from the anterior surface of the forearm. Fibroblasts were obtained as previously described (Iori et al., 2003). For each experiment fibroblasts were thawed and grown in HAM'S F-10 Nutrient Mixture (Sigma Aldrich), supplemented with 10% foetal bovine serum (FBS, Hyclone NL), 4 mM glutamine (Sigma Aldrich), 100 U/mL penicillin and 100 μg/mL streptomycin (Sigma Aldrich) at 37°C until confluence. All studies were performed at the 5-7th passage.

Measurement of Intracellular Calcium

Cells (5×10^5) were seeded onto coverslip (3 cm × 1 cm) and allowed to grow to confluence as described elsewhere (Ceolotto et al., 2001). Briefly, after 24 hours in a quiescent medium without serum, the cells were loaded with 3 μmol/L Fura-2 AM for 1 hour. Then, Fura-2 was removed, a physiological medium containing (mmol/L) NaCl 129, KCl 2.8, KH_2PO_4 0.8, $CaCl_2$ 1, $NaHCO_3$ 8.9, $MgCl_2$ 0.8, glucose 5.6, HEPES 5.6, pH 7.4, was added for 30 min. The coverslip was placed into a quartz cuvette inside a fluorescent spectrophotometer (Shimadzu RF-1501, Kyoto, Japan). A peristaltic perfusion pump delivered solutions at 3 mL/min at the bottom of the cuvette. The baseline fluorescence was obtained by rapidly alternating the excitation wavelength between 340 and 380 nm and recording the 510 nm emission intensity. Intracellular Ca^{2+} concentrations were calculated from the fluorescence ratio recordings as described (Ceolotto et al., 2001). Ang II (100 nM) was added to the perfusion solution when the baseline fluorescence was stable and the fluorescence measurements continued until Ca^{2+}_i concentration recovered to basal level.

Cell Proliferation Rate

Fibroblasts were seeded onto 24 well plates at a density of 2×10^4 cells per well and incubated with HAM-F10 medium containing 10% FBS at 37°C. Then cells were harvested in two wells every day and counted three times each by haemocytometer. Cell proliferation rate was calculated as time required for a cell population to double its number. The doubling time (DT) was calculated in the exponential growth phase by the following equation: $DT= (t_1-t_0) \log_2/(\log N_1-\log N_0)$. Where t_0 is the time at the first counting (day 1), t_1 is the time at the counting on day 5, N_0 is cell number at time t_0 and N_1 is cell number at time t_1. To evaluate the effects of Ang II on proliferation rate, skin fibroblasts were transferred into six well plates at the density of 5×10^4 cells per well and incubated with 10% FBS medium at 37°C. Twenty four hours later, cells in two wells were harvested and counted. The growth medium was replaced, for 6 days, with 3% FBS medium without or with 100 nM Ang II in presence or absence of Losartan, an antagonist of Ang II receptor. On the 6th day, the proliferation rate induced by Ang II was calculated as percentage increase in comparison with control medium.

Chemicals

Angiotensin II, thapsigargin and Fura-2 AM were from Calbiochem (La Jolla, CA). All other chemicals were from Sigma Aldrich (Milan, Italy).

Statistical Analysis

Statistical calculation was carried out using the "Statistical Package for the Social Sciences" program. Differences between group means were tested by analysis of variance, followed by the post-hoc test of Dunnet. The relationship between variables was tested with linear regression analysis. A two-tailed p value of less than 0.05 was considered statistically significant. Data are shown as means ± standard error of the mean (SEM). Experiments were carried on in cell cultures from each subject and performed in duplicate (technical replicates).

Results

Diabetic patients with and without DN had similar ages, sex distribution, duration of diabetes and BMI and required similar doses of insulin. HbA_{1c} levels over six years were similar in patients with DN and in those without as described elsewhere (Ceolotto et al., 2001). Blood pressure was higher in those with nephropathy when compared to those without and normal control subjects. In the group with nephropathy GFR was lower (Table 1), but all patients still had GFR values above 60 mL/min/1.73m^2. Intracellular $Ca^{2+}{}_i$ was higher in quiescent skin fibroblasts from T1DM patients with DN (81±4 nmol/L, $p<0.005$) than in cells from T1DM patients without DN and normal control subjects (64±3 and 60±4, respectively) (Figure 1). Ang II addition evoked a fast and transient rise in $Ca^{2+}{}_i$, which was higher in fibroblasts from T1DM patients with DN (233±17 nmol/L; $p<0.001$) than that observed in cells from T1DM patients without DN (152±9 nmol/L) and in normal subjects (161±2 nmol/L) (Figure 1).

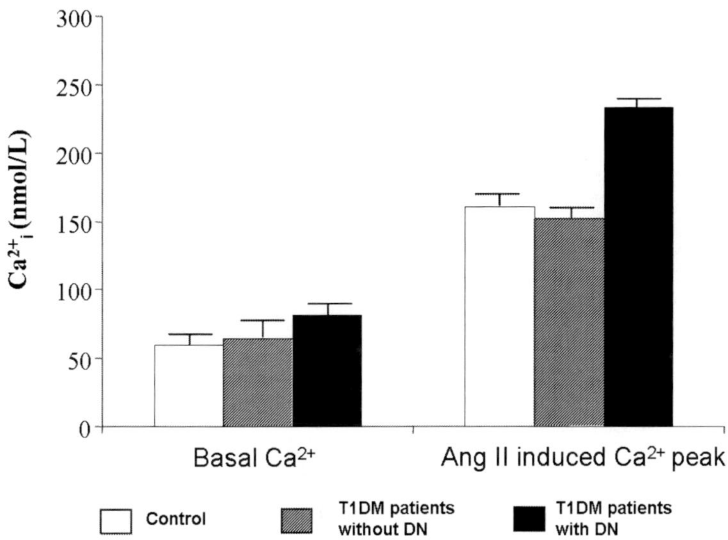

Figure 1. Effect of Ang II on $Ca^{2+}{}_i$ mobilization.

This pattern of response was also present upon chelation of extracellular Ca^{2+} and the Ang II-induced changes of intracellular Ca^{2+} were completely blocked by the specific inhibitor Losartan (100 nmol/L) in all subjects (data not shown). Thapsigargin, a Ca^{2+}-ATPase inhibitor acting on intracellular $Ca^{2+}{}_i$ pools including the one sensitive to *inositol trisphosphate*, has been used to investigate $Ca^{2+}{}_i$ release from intracellular store. $Ca^{2+}{}_i$ release induced by thapsigargin was similar in normal subjects and in T1DM patients with and without DN, suggesting that the different $Ca^{2+}{}_i$ mobilization was unrelated to thapsigargin-sensitive calcium stores (Figure 2). Cell proliferation rate was higher in cells from T1DM patients with DN (mean doubling time 3.8±0.8 days, $p<0.05$) than in cells from T1DM patients without DN and normal controls (mean doubling time 7.8±2 and 7.9±1.9 days, respectively) (Figure 3). In the whole population of diabetic patients, the increase in Ca^{2+}_i after Ang II addition was inversely related to mean doubling time (r=-0.45; p<0.001) (Figure 4). Then, the effects of Ang II on cell proliferation rate in T1DM patients with and without DN and in normal controls were evaluated in fibroblasts grown in 3% FBS medium, as described in the method section. The percentage increase between the proliferation rate evaluated in cells cultured with or without Ang II was significantly higher in T1DM patients with DN (+ 64±10%) in comparison with those without DN (+ 37±9%) and controls (+ 9±5%) (Figure 5).

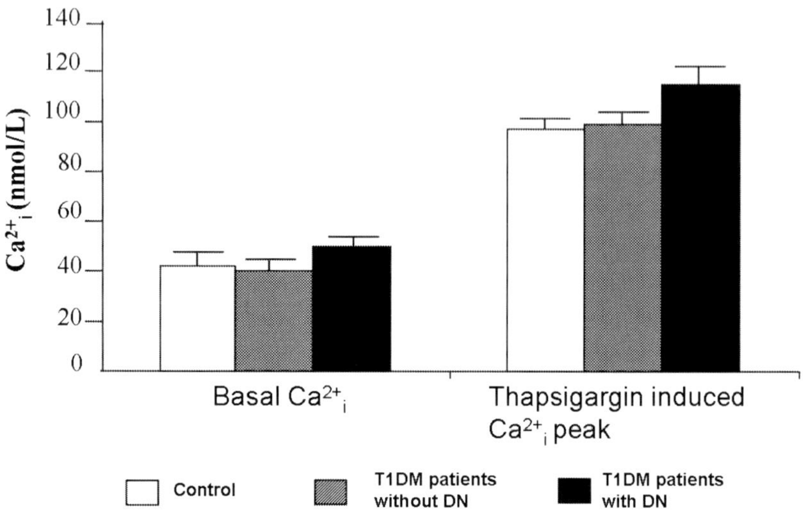

Figure 2. Effect of Thapsigargin on Ca^{2+}_i mobilization. Extracellular Ca^{2+} was chelated by EGTA 1mM.

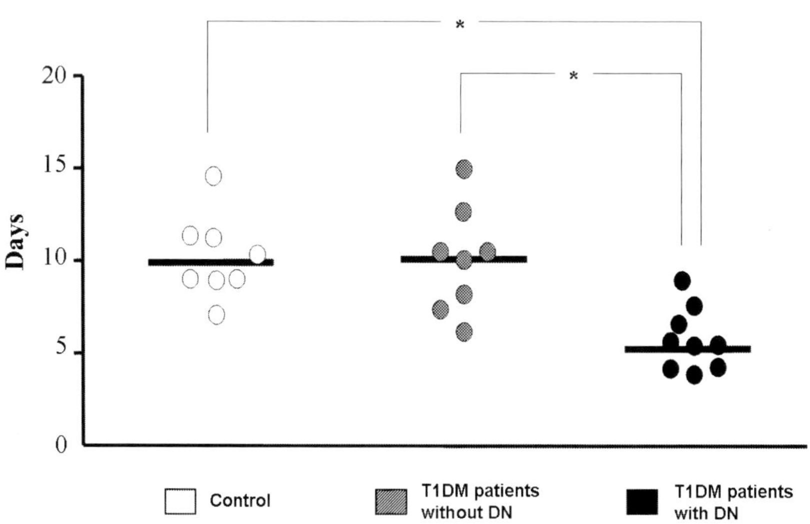

Figure 3. Cell doubling time (days).

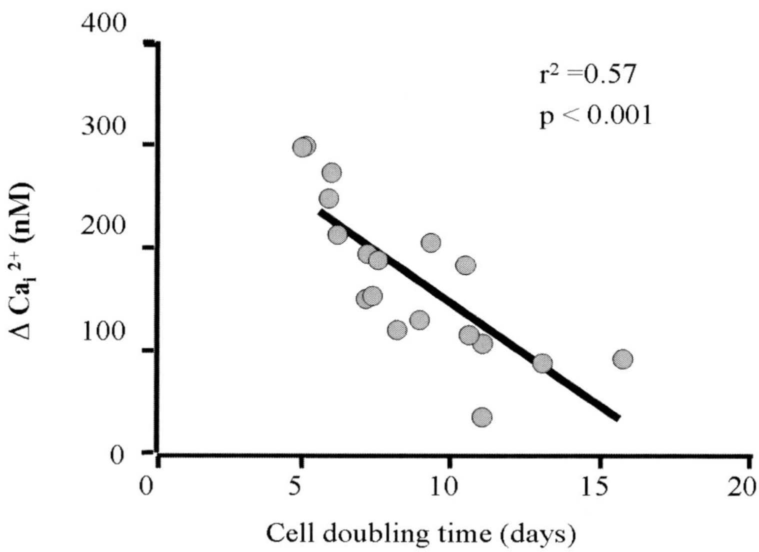

Figure 4. Relationship between Ang II-induced Ca^{2+}_i mobilization and cell proliferation rate.

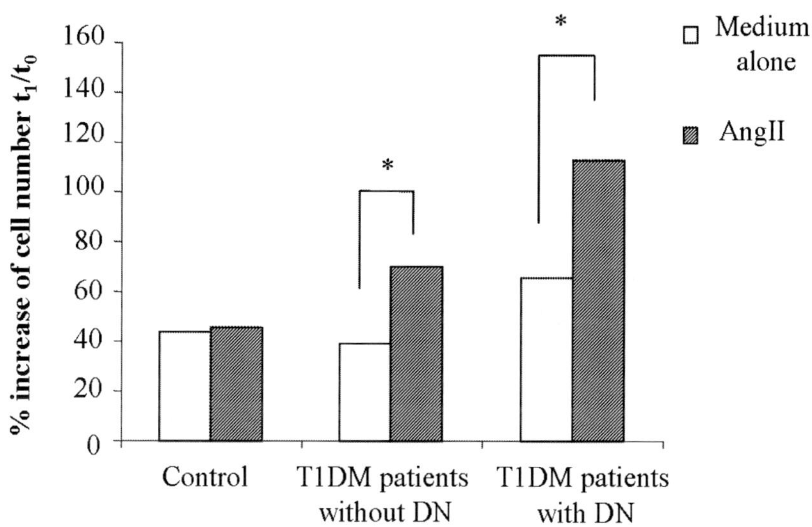

Figure 5. Effect of Ang II on fibroblast proliferation rate.

The difference reached conventional significance only between T1DM patients with DN and normal control subjects ($p<0.05$). All these effects of Ang II on cell proliferation rate were abolished by treatment with Losartan (data not shown).

Discussion

Cultured fibroblasts, obtained from different subjects and grown for several passages with the same growth medium, maintained the characteristic spindle-shaped morphology. In particular, they were strictly diploid and at confluence they formed characteristic parallel arrays and whorls (Ceolotto et al., 1998). Moreover, they synthesized type I collagen (Trevisan et al., 1997-a) and were not positive for factor VIII (Ceolotto et al., 2001). Finally, immunocytochemistry showed a positive staining for the fibroblast-specific marker CD90 in fibroblasts grown up to the 13th passage (Millioni et al., 2010). The specificity of CD90 detection was confirmed by lack of staining upon cells' exposure to the secondary antibody and the detection reagents after omission of the primary antibody. Hence, these results demonstrated the maintenance of a fibroblast phenotype despite prolonged passages in culture of our skin-derived cells. All these evidences, taken together, support the hypothesis that fibroblasts do not undergo phenotypic differentiation after the passage numbers typically used in experiments. This cellular type stability is a great advantage, because it allows to create an experimental setting where the untoward interferences of environmental factors are removed, and the differences observed between cultured fibroblasts from two different clinical conditions, such as T1DM patients with or without DN, could be attributed to a constitutive, genetic predisposition, rather than to the diabetic milieu *per se*. This is particularly important since a genetic predisposition is thought to play a role in the development of this complication, independently from disease duration, metabolic control, or other factors (Ceriello et al., 2000; Huang et al., 2002; Iori et al., 2003; Lurbe, et al., 1996; Schena and Gesualdo, 2005; Trevisan et al., 1992). Alternatively, there is also the possibility that these abnormalities in cells derived from T1DM patients with DN are due to permanent glucose-induced epigenetic changes. Further characterisation of these glucose-induced epigenetic events and the identification of key enzymes involved will improve the authors' understanding of the pathways implicated in diabetic vascular injury. An accelerated sodium–hydrogen antiport (NHE) activity is a typical feature of both essential hypertension and type 1 diabetes (Trevisan and Viberti, 1997_b). This plasma membrane protein catalyzes the exchange of extracellular sodium for intracellular hydrogen and regulates intracellular pH, cell volume and stimulus–response coupling, and cell proliferation. An elevated NHE activity was also reported in immortalized lymphoblasts of T1DM subjects with diabetic nephropathy (Ng et al., 1994) and in cultured skin fibroblasts (Davies et al., 1992; Trevisan et al., 1992). A faster proliferation rate of cultured human skin fibroblasts from patients with T1DM and DN than those without has been described (Trevisan et al., 1992; Lurbe et al., 1996), possibly suggesting a genetic background for this dysfunction. Only a single study contradicts this data, reporting a decreased fibroblast proliferation rate in skin fibroblasts from T1DM patients with DN after the eighth passage (Morocutti et al., 1996). However this divergence could be related to the different culture conditions used in the mentioned study, that is the use of 20% foetal bovine serum (instead of the commonly used 10%) and of high glucose concentration (22 mM instead of 5 mM).

The hypothesis that a faster proliferation rate could be the consequence of an increased NHE activity was abandoned after the demonstration that, in DN and essential hypertension, a faster proliferation rate could not be normalized by the blockade of NHE activity (Lurbe et al., 1996). A possible explanation for this phenomenon could be found in another study (Zeng et al., 2003) where skin fibroblasts from patients with DN exhibited significantly lower mRNA and protein expression of p16, a cyclin-dependent kinase inhibitor which has a key role in cell cycle regulation. In addition, Maestroni et al. (2005) measured skin fibroblasts proliferation rate by cell count in either the presence or the absence of the MAP kinase/ERK kinase (MEK) inhibitor PD 098059. They found that the in vitro inhibition of this key point of the MAP kinase cascade normalized the proliferation rate of skin fibroblasts from T1DM patients with DN, suggesting that the MAP kinase could be involved in this cellular dysfunction. The present study shows an enhanced responsiveness to Ang II in skin fibroblasts from T1DM patients with DN cultured in normal glucose concentrations. In particular Ang II induced $Ca^{2+}{}_i$ mobilization was exaggerated in T1DM patients with DN. This effect was independent of intracellular thapsigargin-sensitive Ca^{2+} store and of extracellular Ca^{2+}. The association between Ang II induced $Ca^{2+}{}_i$ mobilization and cell proliferation rate suggests that an abnormal intracellular signal transduction may be involved in the pathophysiology of DN in Type 1 diabetic patients. We found similar results in fibroblasts from non diabetic patients with essential hypertension (Ceolotto et al., 2001). In those with insulin resistance, we showed that insulin was not able to attenuate Ang II-induced $Ca^{2+}{}_i$ mobilization as in cells from insulin sensitive patients. An abnormality in a pertussis toxin-sensitive G-protein was involved in this altered response. This might be relevant to our study since T1DM patients with DN are also more insulin resistant than those without DN (Yip et al., 1993). Furthermore, immortalised lymphoblasts from Type 1 diabetic patients with nephropathy exhibit increased $Ca^{2+}{}_i$ mobilization after platelet-activating factor stimulation as a consequence of an increased activation of pertussis toxin-sensitive G-protein (Pietruck et al., 1998). Since a predisposition to essential hypertension confers an increased susceptibility to renal damage in diabetic patients is well established (Trevisan and Viberti, 1995) and an altered $Ca^{2+}{}_i$ mobilization was described both in hypertensive patients and in animal models of essential hypertension, our data supports the possibility that an altered $Ca^{2+}{}_i$ metabolism might play a role in the pathogenesis of hypertension in diabetic patients at risk of renal disease.

In vivo hyperglycaemia in diabetic patients is unlikely to explain the results of our study: firstly, HbA_{1c} over the last 6 years before the skin biopsy was similar in T1DM patients with DN and those without DN; secondly, Ang II induced $Ca^{2+}{}_i$ mobilization in cells were virtually identical in patients without DN and in normal control subjects. The effects of the antihypertensive therapy in the diabetic patients with DN, even if they may persist after several *in vitro* cell passages, are also unlikely to explain the differences among the groups. The available evidence from animal studies would suggest that some of the antihypertensive drugs used, particularly the ACE inhibitors, would reduce Ang II activity. Thus, if anything, antihypertensive therapy would tend to reduce rather than enhance the difference. It is also unlikely a role for the impairment in renal function in diabetic patients with nephropathy since no relationship was observed between the decline in GFR and $Ca^{2+}{}_i$ mobilization. In conclusion, all these observations of stable phenotypic changes in cell function, despite serial passages of cells in vitro, suggest a likely intrinsic component in the cell altered response to Ang II in the subset of patients at risk of kidney disease.

References

Caramori, M. L., Fioretto, P., and Mauer, M. (2000). The need for early predictors of diabetic nephropathy risk: is albumin excretion rate sufficient? *Diabetes* 49, 1399-1408.

Ceolotto, G., Pessina, A. C., Iori, E., Monari, A., Trevisan, R., Winkleswski, P., and Semplicini, A. (1998). Modulatory effect of insulin on release of calcium from human fibroblasts by angiotensin II. *Journal of Hypertension* 16, 487-493.

Ceolotto, G., Valente, R., Baritono, E., Reato, S., Iori, E., Monari, A., Trevisan, R., and Semplicini, A. (2001). Effect of insulin and angiotensin II on cell calcium in human skin fibroblasts. *Hypertension* 37, 1486-1491.

Ceriello, A., Morocutti, A., Mercuri, F., Quagliaro, L., Moro, M., Damante, G., and Viberti, G. C. (2000). Defective intracellular antioxidant enzyme production in type 1 diabetic patients with nephropathy. *Diabetes* 49, 2170-2177.

Cooper, M. E. (2001) Interaction of metabolic and haemodynamic factors in mediating experimental diabetic nephropathy. Diabetologia 44, 1957-1972

Davies, J. E., Ng, L. L., Kofoed-Enevoldsen, A., Li, L. K., Earle, K. A., Trevisan, R., and Viberti, G. (1992). Intracellular pH and Na+/H+ antiport activity of cultured skin fibroblasts from diabetics. *Kidney Int.* 42, 1184-1190.

Huang, C., Kim, Y., Caramori, M. L., Fish, A. J., Rich, S. S., Miller, M. E., Russell, G. B., and Mauer, M. (2002). Cellular basis of diabetic nephropathy: II. The transforming growth factorbeta system and diabetic nephropathy lesions in type 1 diabetes. *Diabetes* 51, 3577-3581.

Iori, E., Marescotti, M. C., Vedovato, M., Ceolotto, G., Avogaro, A., Tiengo, A., Del Prato, S., and Trevisan, R. (2003). In situ protein kinase C activity is increased in cultured fibroblasts from type 1 diabetic patients with nephropathy. *Diabetologia* 46, 524-530.

Iori, E., Millioni, R., Puricelli, L., Arrigoni, G., Lenzini, L., Trevisan, R., James, P., Rossi, G. P., Pinna, L. A., and Tessari, P. (2008). Glycolytic enzyme expression and pyruvate kinase activity in cultured fibroblasts from type 1 diabetic patients with and without nephropathy. *Biochimica. et Biophysica. Acta.* 1782, 627-633.

Ismail, N., Beckner, B., Strtzelczyk, P., and Ritz, E. (1999). Renal disease and hypertension in non-insulin-dependent diabetes mellitus. *Kidney International* 55, 1-28.

Lam, S., van der Geest, R. N., Verhagen, N. A., Daha, M. R., and van Kooten C. (2004). Secretion of collagen type IV by human renal fibroblasts is increased by high glucose via a TGF-beta-independent pathway. *Nephrology, Dialysis, Transplantation* 19, 1694-1701.

Leehey, D. J., Singh, A. K., Alavi, N., and Singh, R. (2000). Role of angiotensin II in diabetic nephropathy. *Kidney Int. Suppl.* 77, S93-S98.

Lurbe, A., Fioretto, P., Mauer, M., LaPointe, M. S., and Batlle, D. (1996). Growth phenotype of cultured skin fibroblasts from patients with insulin-dependent diabetes mellitus with and without nephropathy and its association to overactivity of the Na^+/H^+ antiporter. *Kidney International* 50, 1684-1693.

Maestroni, A., Tentori, F., Meregalli, G., Gabellini, D., Asnaghi, V., Ruggieri, D., and Zerbini, G. (2005). Inhibition of MAP-kinase cascade normalizes the proliferation rate of fibroblasts from patients with Type 1 diabetes and nephropathy. *Journal of Diabetes Complications* 19, 291-296.

Millioni, R., Iori, E., Lenzini, L., Puricelli, L., Caroccia, B., Arrigoni, G., Rossi, G. P., and Tessari P. (2010). Caldesmon over-expression in type 1 diabetic nephropathy. *J. Diabetes Complications* 25(2), 114-121.

Millioni, R., Iori, E., Puricelli, L., Arrigoni, G., Vedovato, M., Trevisan, R., James, P., Tiengo, A., and Tessari, P. (2008). Abnormal cytoskeletal protein expression in cultured skin fibroblasts from type 1 diabetes mellitus patients with nephropathy: a proteomic approach. *Proteomics Clinical Applications* 2(4), 492-503.

Morocutti, A., Earle, K. A., Sethi, M., Piras, G., Pal, K., Richards, D., Rodemann, P., and Viberti, G. (1996). Premature senescence of skin fibroblasts from insulin-dependent diabetic patients with kidney disease. *Kidney Int.* 50, 250-256.

Nathan, D. M., Zinman, B., Cleary, P. A., Backlund, J. Y., Genuth, S., Miller, R., Orchard, T. J. (2009). Modern-day clinical course of type 1 diabetes mellitus after 30 years' duration: the diabetes control and complications trial/epidemiology of diabetes interventions and complications and Pittsburgh epidemiology of diabetes complications experience (1983–2005). *Archives of Internal Medicine* 169, 1307-1316.

Najafian, B., and Mauer, M. (2009). Progression of diabetic nephropathy in type 1 diabetic patients. *Diabetes Research and Clinical Practice*. 83(1), 1-8.

Ng, L. L, Davies, J. E., Siczkowski, M., Sweeney, F. P., Quinn, P. A., Krolewski, B., Krolewski, A. S. (1994). Abnormal Na^{+}/H^{+} antiporter phenotype and turnover of immortalized lymphoblasts from type 1 diabetic patients with nephropathy. *Journal of Clinical Investigation* 93, 2750-2757.

Pietruck, F., Spleiter, S., Daul, A., Philipp, T., Derwahl, M., Schatz, H., and Siffert, W. (1998). Enhanced G protein activation in IDDM patients with diabetic nephropathy. *Diabetologia* 41, 94-100.

Podestà, F., Meregalli, G., Ghelardi, R., Del Giudice, R., Asnaghi, V., Maestroni, A., and Zerbini, G. (2001). Low Ca^{2+} pump activity in diabetic nephropathy. *American Journal of Kidney Disease* 38, 465-472.

Puricelli, L., Iori, E., Millioni, R., Arrigoni, G., James, P., Vedovato, M., and Tessari, P. (2006). Proteome analysis of cultured fibroblasts from type 1 diabetic patients and normal subjects. *The Journal of Clinical Endocrinology and Metabolism* 91, 3507-3514.

Schena, F. P., and Gesualdo, L. (2005). Pathogenetic mechanisms of diabetic nephropathy. *Journal of American Society of Nephrology* 16, 30-33.

Singh, R., Alavi, N., Singh, A. K., and Leehey, D. J. (1999). Role of angiotensin II in glucose-induced inhibition of mesangial matrix degradation. *Diabetes* 48, 2066-2073.

Singh, R., Singh, A. K., Alavi, N., and Leehey, D. J. (2003). Mechanism of increased angiotensin II levels in glomerular mesangial cells cultured in high glucose. *Journal of American Society of Nephrology* 14(4), 873-80.

Steffes, M. W., Bilous, R. W., Sutherland, D. E., and Mauer, S. M. (1992). Cell and matrix components of the glomerular mesangium in type I diabetes. *Diabetes* 41, 679-684.

Tessari, P., Puricelli, L., Iori, E., Arrigoni, G., Vedovato, M., James, P., Coracina, A., and Millioni, R. (2007). Altered chaperone and protein turnover regulators expression in cultured skin fibroblasts from type 1 diabetes mellitus with nephropathy. *Journal of Proteome Research* 6, 976-986.

Toyoda, M., Najafian, B., Kim, Y., Caramori, M. L., and Mauer, M. (2007). Podocyte detachment and reduced glomerular capillary endothelial fenestration in human type 1 diabetic nephropathy. *Diabetes* 56, 2155-2160.

Trevisan, R., Li, L., Messent, J., Tariq, T., Earle, K., Walker, J. D., and Viberti, G. (1992). Na^{+}/H^{+} antiport activity and cell growth in cultured skin fibroblasts of IDDM patients with nephropathy. *Diabetes* 41,1239-1246.

Trevisan, R., and Viberti, G. (1995). Genetic factors in the development of diabetic nephropathy. *The Journal of Laboratory and Clinical Medicine* 126, 342-349.

Trevisan, R., and Viberti, G. (1997-b). Sodium-hydrogen antiporter: its possible role in the genesis of diabetic nephropathy. *Nephrology, Dialysis, Transplantation* 12, 643-645.

Trevisan, R., Yip, J., Sarika, L., Li, L. K., and Viberti G. (1997-a). Enhanced collagen synthesis in cultured skin fibroblasts from insulindependent diabetic patients with nephropathy. *Journal of The American Society of Nephrology* 8, 1133-1139.

Yip, J., Mattock, M. B., Morocutti, A., Sethi, M., Trevisan, R., and Viberti, G. (1993). Insulin resistance in insulin-dependent diabetic patients with microalbuminuria. *Lancet* 342, 883-887.

Zeng, L., Kanwar, Y. S., Amro, N., Phillips, C., Molitch, M., Batlle, D., and Danesh, F. R. (2003). Epigenetic and genetic analysis of p16 in dermal fibroblasts from type 1 diabetic patients with nephropathy. *Kidney International* 63(6), 2094-2102.

In: Dermal Fibroblasts
Editor: Xiaowen Bai
ISBN: 978-1-62618-799-3

Chapter 3

Skin Connective Tissue Aging and Dermal Fibroblasts

Taihao Quan *
Department of Dermatology, University of Michigan
Medical School, Ann Arbor, Michigan, US

Abstract

The bulk of human skin is composed of a dense collagen-rich connective tissue, which provides structural and mechanical support of the skin. The collagen-rich connective tissue is produced, organized, and maintained by fibroblasts in the dermis. By creating dermal connective tissue structural and mechanical environment, dermal fibroblasts control skin's homeostasis. The major task of the skin is to protect the human body from harmful environmental factors, such as solar ultraviolet (UV) radiation and micro-organisms. During aging, the dermal connective tissue collagen, the most abundant protein in skin, undergoes progressive alterations, which cause skin connective tissue aging. Age-related alteration of dermal collagen reduces skin dermis structural and mechanical properties and creates a tissue microenvironment that promotes age-related skin diseases, such as poor wound healing and skin cancer. This chapter describes the role of dermal fibroblasts in skin connective tissue aging and in age-related skin diseases.

Keywords: Dermal fibroblast, ECM, collagen, connective tissue, stromal microenvironment, mechanical property, skin connective tissue aging, age-related diseases

Introduction

Human skin is the largest and the heaviest organ of the human body. The primary function of the skin is to provide a barrier to external insults, such as wounding, solar UV,

* Corresponding author: Taihao Quan. Address: 1301 Catherine, Med Sic I, Room 6435, Ann Arbor, MI 48109; E-mail: thquan@umich.edu.

and micro-organisms. (Chuong et al., 2002; Farage et al., 2010). The human skin is composed of three layers: the epidermis (outermost layer), the dermis (vascular connective tissue below the epidermis) and the hypodermis (the deepest portion of the skin) (Bolognia et al., 2008; Farage et al., 2010). The epidermis is primarily composed of keratinocytes. Keratinocytes produce a class of proteins known as keratins that are a major component of the protective horny layer. The dermis is composed of a dense connective tissue such as collagen-rich extracellular matrix (ECM) and dermal fibroblasts. The subcutaneous hypodermis consists of fat cells dispersed throughout the connective tissue framework. The dermis intimately interacts with the epidermis and subcutaneous fat in maintaining the structural and mechanical properties of the skin. The dermis is less cellular than the epidermis and is composed predominantly of ECM proteins, such as collagen, elastin, fibronectin, and proteoglycans. As human skin is the largest organ of the body, dermal collagen represents by far the most abundant protein that constitutes the bulk (90% dry weight) of skin (Bernstein et al., 1996; Uitto, 1986). Dermal connective tissue collagen is essentially responsible for the skin's tensile strength and mechanical properties. In human skin dermis, collagen-rich ECM is synthesized, organized, and maintained by dermal fibroblasts (Farage et al., 2010; Fisher et al., 2008). Dermal fibroblasts secrete the precursors of all the components of the ECM proteins in skin. Therefore, the main function of dermal fibroblasts is to maintain the structural and mechanical integrity of dermal connective tissues by continuously synthesizing and secreting of collagen and other ECM proteins. By creating dermal ECM connective tissue, dermal fibroblasts provide structural and mechanical support to skin and thus control skin's homeostasis including maintenance of epidermal barrier/protective functions and regulation of body temperature, sensory reception, vascularization, and water balance. Therefore, collagen is the principal structural protein in skin, and alteration of dermal fibroblast function has significant impact on skin function. Dermal fibroblasts are derived from primitive mesenchyme (Shamis et al., 2011). Thus, dermal fibroblasts express the intermediate filament protein vimentin, a marker to distinguish their mesodermal origin. Under nonpathologic conditions, dermal fibroblasts are quiescent and less likely to replicate. However, under pathological conditions such as upon wound, dermal fibroblasts undergo transition from a quiescent to an activated state. At this point, dermal fibroblasts proliferation can stimulate and also can differentiate into myofibroblasts, a type of fibroblast cells with smooth muscle characteristics by expression of alpha-smooth muscle actin (α-SMA). Myofibroblasts have strong contractile property, which plays a critical role in wound healing by pulling dermal ECM (Gabbiani, 2003).

Collagen is not only the main component of the skin, but also is the most abundant protein, making up about 25% to 35% of the whole-body protein content (Di Lullo et al., 2002). Collagen is mostly found in fibrous connective tissue such as tendon, ligament and skin, and is also abundant in cornea, cartilage, bone, blood vessels, the gut, and intervertebral disc. The matured collagen composition is particularly rich in four amino acids: lysine, proline, hydroxylysine and hydroxyproline (Szpak, 2011). The most common patterns in the amino acid sequence of collagen are lysine-hydroxylysine-proline and lysine-hydroxylysine-hydroxyproline. Collagens are synthesized and secreted by fibroblasts in different forms known as types. It is important that fibroblasts must synthesis the right amount and the right type of collagen in the right place. In human skin, type I and type III collagens are the most abundant types of collagen. Other types of collagen in the skin are collagen V, VI, and XII, which are found in much smaller amounts and appear to have a supportive role. Type I collagen consists of three polypeptide chains organized in a triple helical configuration, which consists of two identical α1 chains and an additional α2 chain (Perumal et al., 2008; Szpak, 2011). The fibroblast produces soluble triple helix collagen precursor known as procollagen. Each procollagen polypeptide chain is originally synthesized with additional amino acids at both ends that confer solubility. Procollagen is enzymatically-cleaved in the extracellular space. Removal of the end domains produces collagen, which spontaneously assembles into large fibers that are cross-linked by the enzyme lysyl oxidase (Rodriguez et al., 2008). This cross-linking is necessary to stabilize mature collagen fibers, which provide structural and mechanical properties to skin. Because procollagens and their partially processed forms are precursor molecules for mature collagen, their levels generally reflect the level of collagen biosynthesis.

Type I collagen, like all proteins, undergoes natural breakdown by enzymatic degradation; however, this degradation in human skin is exceedingly slow, with calculated half-life of 17 years (Verzijl et al., 2000). Humans express several enzymes that are capable of initiating breakdown of type I collagen. These enzymes are members of a family of matrix protein-degrading enzymes, referred to as matrix metalloproteinases (MMPs), as discussed in more detail below (Hegedus et al., 2008; Lapiere Ch, 2005). In human skin, dermal fibroblasts are the major source of MMPs production. MMPs comprise a large family of proteinases that are capable of degrading every type of dermal ECM protein (Page-McCaw et al., 2007). To date, the MMP gene family consists of 25 members, 24 of which are expressed in mammals (Egeblad and Werb, 2002; Hegedus et al., 2008). In human skin, the transcripts for MMP-8, -10, -12, -20, and -26 are not detected (Quan et al., 2009). MMP-14, -2, -3, -28, -7, and -15 are the most highly expressed, while remaining MMPs are expressed at lower levels. MMP-1, appears to be primary enzyme involved in normal turnover of skin collagen (Brennan et al., 2003; Fisher et al., 1996b; Fisher et al., 1997a). In healthy young skin, MMP-1 expression is exceedingly low, near the limit of detection by the RT-PCR (Fisher et al., 2009; Quan et al., 2009). However, MMP-1 is potently induced by wide range of stimulus such as UV irradiation and reactive oxygen species (Fisher et al., 2009; Quan et al., 2009). *After* primary *cleavage* of *collagen* at a single site by *MMP-1*, collagen denatures, and then undergoes further degradation by other members of the MMP family (Brennan et al., 2003; Fisher et al., 1996a; Fisher et al., 1997b). In addition, skin expresses natural MMP inhibitors, including tissue inhibitors of matrix metalloproteinases (TIMPs) and the proteoglycan decorin. These inhibitors further act to retard collagen breakdown (Gomez et al., 1997).

Although dermal collagen in human skin is stable, dermal collagen undergoes progressive gradual alterations and accumulations of damage through over years or decades, which eventually cause skin connective tissue aging seen in aged skin.

Human Skin Connective Tissue Aging and Dermal Fibroblasts

Human skin, like all other organs, undergoes natural aging process as a consequence of the passage of time.

Meantime, human skin, unlike other organs, continuously experience harmful stress and damage from environmental sources such as solar ultraviolet (UV) radiation and micro-organisms. (Fisher et al., 2008; Yaar and Gilchrest, 1998). Ultraviolet irradiation from the sun is a well-recognized, potent environmental insult capable of damaging skin tissue. UV-induced skin damage includes sunburn, immune suppression, cancer, and premature skin aging (photoaging) (Biesalski et al., 2003; Fisher et al., 2002; Taylor and Sober, 1996; Yaar and Gilchrest, 2007).

Based on its causes, cutaneous aging can be classified into two types: natural aging, also known as intrinsic aging, and photoaging, also known as extrinsic aging. Natural aging refers to those changes observed in all individuals resulting from the passage of time, whereas photoaging refers to those changes attributable to habitual sun exposure. Both processes are cumulative and therefore photoaging is superimposed on intrinsic aging. Therefore, the alterations seen in aged skin are a combination of intrinsic and extrinsic aging. The most clinically noticeable age-related changes occur on face, neck, forearm, and lower leg skin (Wlaschek et al., 2001). These areas undergo a combination of natural aging and photoaging and age-related skin diseases occur most *often* in these areas. Aging skin is recognizable by fine and course wrinkles, skin laxity, coarseness, uneven pigmentation, and brown spots.

Histological and ultrastructural studies have revealed that the major alterations in photoaged skin are localized in the connective tissue dermis (Fisher et al., 1996b; Fisher et al., 1997a). Biochemical evidence of connective tissue damage in aged human skin includes alterations of collagen and elastin. Since collagen fibrils and elastin are responsible for the strength and resiliency of skin, their degeneration with aging causes skin to become fragile and easily bruised. Histologically, connective tissue damage induced by ultraviolet irradiation manifests as a disorganization of collagen fibrils, massive accumulation of abnormal, amorphous, and elastin-containing material. The accumulation of elastotic material is accompanied by degeneration of the surrounding collagen matrix. Naturally-aged skin differs from photoaged skin.

The severity of changes in the dermal matrix of naturally-aged skin relative to photoaged skin is less. Only a modest increase in the number and thickness of elastic fibers is observed. However, reflecting the general atrophic state of naturally-aged skin, there are reduced collagen production and increased collagen damage, compared to younger skin (Fisher et al., 2009). Figure1 shows nanoscale topographical images of dermal collagen fibrils assessed by atomic force microscopy (AFM).

In young (20-30 years) or sun-protected underarm (50-60 years) human dermis, intact collagen fibrils are abundant, tightly packed, and well-organized (left panels). In contrast, collagen fibrils are damaged and disorganized in aged (80 +years) or sun-exposed forearm (50-60 years) human dermis (right panels). Therefore, in general, naturally-aged and photoaged skin share common molecular features, characterized by aberrant collagen homeostasis that includes reduced production of collagen, and increased collagen degradation (Fisher et al., 2000; Fisher et al., 2002; Fisher et al., 2009; Fisher et al., 2008; Quan et al., 2010b).

This aberrant collagen homeostasis results in net collagen deficiency which is characteristic features of skin connective tissue aging. Age-related aberrant collagen homeostasis impairs the structural integrity and mechanical properties of the skin, and causes age-related skin diseases. Therefore, aberrant collagen homeostasis is central event in human skin connective tissue aging, which is virtually caused by impaired dermal fibroblasts function, as described below.

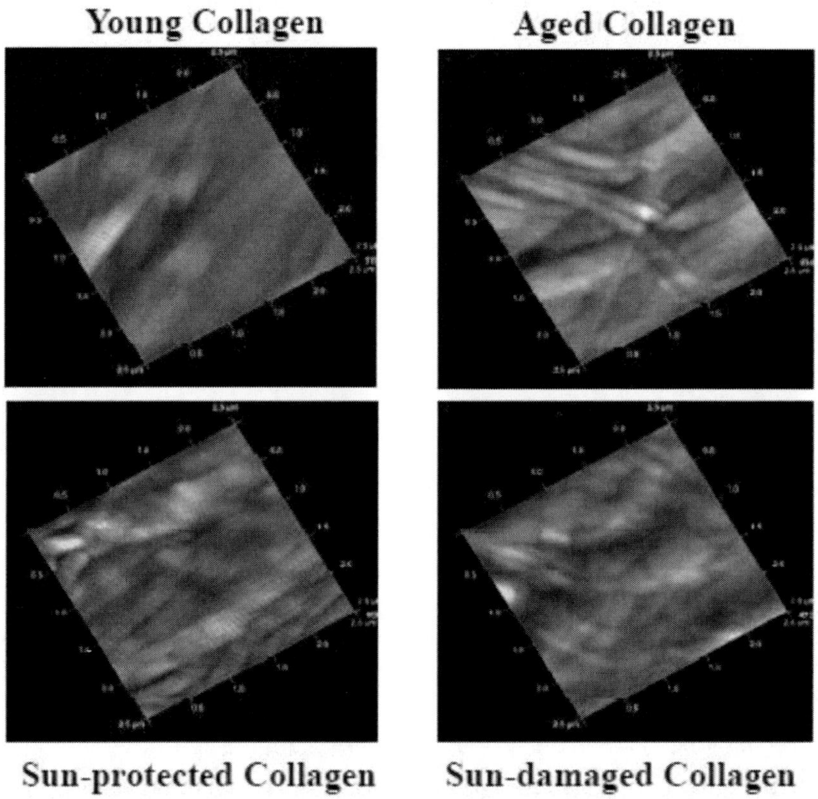

Figure 1. Alterations of dermal connective tissue collagen are characteristic features of human skin natural aging and photoaging. Nanoscale human skin dermal collagen fibrils are analyzed by atomic force microscopy (AFM). Representative AFM images of dermal collagen fibrils in young (22 years)/aged (84 years) and sun-protected underarm/sun-damaged forearm skin (52 years).

Impaired Transforming Growth Factor-β (TGF-β) Pathway in Dermal Fibroblasts and Skin Connective Tissue Aging

A wealth of evidence indicates that TGF-β plays a central role in ECM biosynthesis and is critical for tissue remodeling during wound healing (Ignotz and Massague, 1986; Patil et al., 2011; Varga et al., 1987). In human dermal fibroblasts, TGF-β not only functions as primary regulator for collagen synthesis, but also prevents collagen degradation by inhibiting MMP-1 expression (Hall et al., 2003; Quan et al., 2002a). Therefore, alterations of TGF-β signaling pathway have significant impact on collagen homeostasis in human skin. In mammals, three isoforms of TGF-β, TGF-β1, TGF-β2, TGF-β3, have been identified (Massague, 1990). TGF-β initiates its cellular action by binding to its cell surface receptor complex, composed of TGF-β type I receptor (TβRI) and TGF-β type II receptor (TβRII) (Massague, 2000). Binding of TGF-β allows TβRII to activate TβRI, which in turn activates transcription factors, Smad2 and Smad3. Activated Smad2 or Smad3 forms heteromeric complexes with the common partner, Smad4. These activated Smad complexes translocate into nucleus, where they interact with Smad Binding Element (SBE) in the promoter regions of TGF-β target genes, such as collagen, connective tissue growth factor (CCN2/CTGF), fibronectin, and MMP-1.

Since deregulation of collagen homeostasis is a prominent feature in aged human skin and TGF-β pathway plays a significant role in collagen homeostasis, TGF-β signaling has been investigated in human skin (Quan et al., 2002a; Quan et al., 2002b; Quan et al., 2004; Quan et al., 2001; Quan et al., 2010b). Laser capture microdissection (LCM) coupled quantitative real-time RT-PCR indicates that TβRII mRNA level is reduced 59% in aged (80+years) skin dermis, compared to young (20-30 years) skin dermis *in vivo* (Quan et al., 2006; Quan et al., 2010b).

Similarly, TβRII mRNA level is reduced 38% in photodamaged forearm skin, compared with subject-matched, sun-protected hip skin (Quan et al., 2006).

In contrast to TβRII, TβRI mRNA levels are similar in young and aged or photodamaged and sun-protected human skin *in vivo*. Interestingly, UV irradiation of human skin results in substantial reduction of TβRII gene expression (Quan et al., 2004), as seen in naturally-aged and photoaged human skin *in vivo*. In contrast, TβRI gene expression is unaffected by UV irradiation, as observed in naturally-aged and photoaged human skin *in vivo*, suggesting that UV irradiation specifically reduces TβRII. UV irradiation of cultured adult human dermal fibroblasts indicates that both TβRII mRNA and protein levels are significantly decreased, while UV irradiation had no effect on expression of either TβRI mRNA or protein (Quan et al., 2004). As described above, the actions of TGF-β are initiated in response to binding of TGF-β it its cell surface receptor complex. Reduced expression of TβRII causes decreased binding of TGF-β to the surface of dermal fibroblasts (Quan et al., 2001). This loss of TGF-β binding results in decreased cellular responsiveness to TGF-β (Quan et al., 2001). In the absence of functional receptor-binding, TGF-β is no longer able to stimulate phosphorylation of its intracellular effectors Smad2 and Smad3.

Therefore, Smad2 and Smad3 do not translocate into the nucleus or alter transcription of target genes. Electrophoretic Mobility Shift Assays (EMSA), using the Smad 3 Binding Element (SBE) as probe, shows that exposure of human skin to UV irradiation substantially reduces formation of the TGF-β-induced DNA/Smad complex (Quan et al., 2002b). Preventing loss of TβRII by over-expression restores TGF-β activation of Smad3, and therefore protects against UV inhibition of type I procollagen gene expression (Quan et al., 2004). These data provide strong support for the concept that down-regulation of TβRII is a critical event in reduced procollagen synthesis in naturally-aged and photoaged human skin.

As described above, Smad2, Smad3, and Smad4 function to transduce the TGF-β signal from the cell surface into the nucleus and regulate transcription of TGF-β target genes.

In contrast, Smad7 functions to interfere with TGF-β-induced activation of Smad2 and Smad3, and thereby acts to block TGF-β signaling. Expression of Smad7 is inducible by TGF-β itself, indicating a negative feedback mechanism that regulates TGF-β/Smad signaling (Afrakhte et al., 1998; Hayashi et al., 1997; Itoh et al., 1998; Nakao et al., 1997).

Aberrant expression of Smad7 could disturb the delicate balance inherent in TGF-β/Smad signaling, and thereby contribute to pathophysiology of certain human diseases (Fiocchi, 2001; He et al., 2002; Nakao et al., 2002; Varga, 2002).

For example, in the fibrotic disease scleroderma, the level of Smad7 expression has been reported to be reduced (Dong et al., 2002). This reduction could permit unrestrained activation of the TGF-β/Smad pathway, resulting in excess production and deposition of collagen, the hallmark of fibrosis. Conversely, Smad7 has been reported to be over-expressed in inflammatory bowel disease (Monteleone et al., 2001). This over-expression could down regulate TGF-β-mediated immunosuppression, resulting in hyper-secretion of proinflammatory cytokines. Smad2, Smad3 and Smad4 are constitutively expressed in human skin and cultured human dermal fibroblasts, and that UV irradiation has no significant effect on the protein level of any of these three Smad proteins (Quan et al., 2005).

In contrast, UV irradiation rapidly up-regulates expression of Smad7 mRNA and protein in human skin *in vivo* and cultured human dermal fibroblasts (Quan et al., 2005). Importantly, this induction of Smad7 by UV irradiation results inhibition of TGF-β-induced type I procollagen gene expression. UV irradiation potently induces Smad7 promoter reporter activity, indicating that Smad7 induction is, at least in part, due to increased transcription. The 5'-flanking region of the Smad7 promoter contains binding sequences for transcription factors Smad3 and AP-1 (Brodin et al., 2000). These regulatory elements and flanking sequences of the Smad7 gene promoter are near completely conserved in human, rat and murine genes (Stopa et al., 2000). Transcription factors Smad3 and AP-1 are necessary for optimal induction of Smad7 transcription (Brodin et al., 2000; Nagarajan et al., 1999; von Gersdorff et al., 2000). Disruption of Smad3 element abolishes TGF-β inducibility, but does not greatly impair basal expression (Quan et al., 2005). Disruption of AP-1 binding element substantially impairs basal and TGF-β-stimulated Smad7 expression. UV irradiation reduces protein binding to the Smad3 site, and increases binding to the AP-1 site. Analysis of proteins bound to the AP-1 site reveals increased binding of AP-1 family members, c-Jun and c-Fos. Deletion of AP-1 binding site in the Smad7 promoter completely abolishes UV stimulation of Smad7 transcription. Furthermore, over-expression of dominant negative c-Jun substantially reduces UV induction of Smad7 transcription (Quan et al., 2005). These data demonstrate that UV irradiation induces Smad7 via activation of transcription factor AP-1 in human dermal fibroblasts. UV irradiation activates multiple cell surface growth factor and cytokine receptors, which initiates intracellular signal transduction cascades (Fisher et al., 1996b; Fisher et al., 2002; Fisher et al., 1997a; Rittie and Fisher, 2002). These activated signaling cascades activate target transcription factors, including AP-1. These data strongly suggest that UV-activated AP-1 stimulates Smad7 expression, which impairs TGF-β responsiveness, the primary regulator of type I collagen. Furthermore, Type I (α2) procollagen promoter activity is reduced by UV, and further reduced by overexpression of Smad7 (Quan et al., 2001), suggesting that UV inhibition of procollagen gene expression is mediated, at least in part, by induction of Smad7. As relates to UV inhibition of procollagen biosynthesis, a single exposure to UV irradiation (2MED) causes a near-complete loss of procollagen synthesis which persists for 24 hours, followed by recovery 48-72 hours post UV in human skin *in vivo* (Fisher et al., 1998). In non-irradiated human skin, TGF-β2 is more abundant than TGF-β1 or TGF-β3 (Quan et al., 2002b). Following UV irradiation, there is a rapid and sharp decrease in TGF-β2 that coincides with a UV-induced loss of procollagen synthesis within 24 hours post UV (Quan et al., 2002b). TGF-β1 and TGF-β3 show a gradual increase, starting at 24 hours post UV that coincides with recovery of procollagen synthesis, observed at 48-72 hours post UV.

For these reasons, it is postulated that UV-induced reductions in collagen synthesis in human skin *in vivo* could be caused, at least in part, by impairment of TGF-β/Smad signaling pathway. In naturally-aged human skin dermis, TGF-β1 mRNA and protein expression are markedly reduced, compared to young skin dermis (Quan et al., 2010b). Laser capture microdissection confirmed that TGF-β1 mRNA in captured fibroblasts is significantly reduced in aged dermal fibroblasts, compared to young dermal fibroblasts in human skin *in vivo*. Above data provide strong evidence that impairment of TGF-β/Smad signaling in dermal fibroblasts largely contributes to aberrant collagen homeostasis and skin connective tissue aging.

Elevated Matrix-Degrading Metalloproteinases (MMPs) in Dermal Fibroblasts and Skin Connective Tissue Aging

Human skin connective tissue aging is caused by two primary pathways, either by inhibiting procollagen biosynthesis or by stimulating collagen breakdown, resulting in reduced or disorganized collagen fibrils in skin dermis (Fisher et al., 1998; Fisher et al., 1997a; Quan et al., 2010b). Matrix-degrading metalloproteinases (MMPs), which degrade collagen and other ECM proteins, are elevated in naturally-aged and photoaged and UV-irradiated human skin *in vivo* (Fisher et al., 2009; Fisher et al., 1997a; Quan et al., 2009). Elevated MMPs over years or decades is considered to be the major contributor to the skin connective tissue damage seen in naturally-aged and photoaged skin. MMPs comprise a family of zinc-containing proteinases (Stamenkovic, 2003; Sternlicht and Werb, 2001). MMPs are classified as collagenases, gelatinases, stromelysins, and membrane-type MMPs according to their substrate specificities and whether they are secreted soluble proteins or bound to cell surface membrane (Fu et al., 2008). MMPs are involved in a wide range of proteolytic events in physiological and pathological circumstances including embryogenesis, wound healing, inflammation, angiogenesis, and cancer (Egeblad and Werb, 2002; Kerkela and Saarialho-Kere, 2003; Sternlicht and Werb, 2001).

Studies conducted by us and others over the past several years have revealed that UV radiation elevates at least three different MMPs in human skin *in vivo*, i.e., interstitial collagenase (MMP-1), stromelysin-1 (MMP-3), and 92kDa gelatinase (MMP-9) (Brenneisen et al., 2002; Quan et al., 2009). MMP-1 and MMP-3 mRNA levels are induced several thousand-fold at 24 hours post UV irradiation, whereas MMP-9 is modestly induced in human skin *in vivo*. MMP-1 initiates cleavage of fibrillar collagen, typically type I and III in skin, at a single site within its central triple helix. Once cleaved by MMP-1, collagen can be further degraded by elevated levels of MMP-3 and MMP-9 (Brennan et al., 2003; Fisher et al., 1996b). The combined actions of MMP-1, 3, and 9 have the capacity to degrade most of the proteins that comprise the dermal ECM. Exposure of sun-protected human skin to purified human MMP-1 in organ culture causes collagen fragmentation, and alterations in the structure and organization of collagen fibrils in the dermis that resemble those observed in naturally-aged and photoaged skin (Fisher et al., 2009; Fligiel et al., 2003; Varani et al., 2001).

These studies suggest that MMP-1, MMP-3, and MMP-9 are primary UV-inducible collagenolytic enzymes, and MMP-1 is the major protease capable of initiating degradation of native fibrillar collagens in human skin *in vivo*. Laser capture microdissection (LCM) coupled with real-time RT-PCR indicates that all three UV irradiation inducible MMPs, are primarily produced by the epidermis, rather than dermis in human skin *in vivo* (Quan et al., 2009). Epidermal keratinocytes are the major cellular source of MMPs that are produced in response to exposure of human skin to solar UV irradiation.

Although collagen-degrading MMP-1, 3 and 9 are primarily induced in the epidermis, the secreted enzymes diffuse into the dermis and degrade collagen. However, it is possible that dermal cells may also play a role in epidermal production of MMPs, through indirect paracrine mechanisms, by release of growth factors or cytokines, which in turn modulate MMP production by epidermal keratinocytes. It is also noteworthy that solar UV radiation readily penetrates the reticular dermis, rendering dermal fibroblasts accessible targets (Farage et al., 2008; Wlaschek et al., 2001). As dermal fibroblasts are the major cell type of the dermal compartment, UV-inducible collagenolytic enzymes are also induced by dermal stromal cell including fibroblasts (Brenneisen et al., 2002).

Compared to acute UV irradiation, a larger variety of MMPs, including UV-inducible MMPs, are constitutively elevated in chronic photodamaged skin. Interestingly, compared to acute UV irradiation, in which the epidermis is the major source of transiently induced MMPs (Quan et al., 2009), the dermis is the major source of elevated MMPs in photodamaged skin (unpublished data). Elevated MMPs in photodamaged dermis can be divided into following groups: collagenases, MMP-1; gelatinases, MMP-2 and MMP-9; stromelysins, MMP-3 and MMP-11; membrane-associated: MMP-17, and recently identified MMP-27. Additionally, compared to acute UV irradiation, in which TIMP-1 is significantly induced (Fisher et al., 1997a), elevated MMPs in photoaged skin are not accompanied by alterations of TIMPs expression. Similarly, multiple MMPs are elevated in naturally-aged human skin dermis other than epidermis (unpublished data), suggesting the combined actions of the wide variety of MMPs that are constitutively elevated in naturally-aged and photodamaged dermis are involved in progressive degradation of dermal ECM.

MMPs are strongly regulated by transcription factor AP-1, which is rapidly induced and activated by UV irradiation in human skin *in vivo* (Birkedal-Hansen et al., 1993; Fisher et al., 1996b; Mauviel, 1993). UV irradiation rapidly up-regulates c-Jun expression by activating the mitogen activated protein (MAP) kinase pathway. Elevated levels of c-Jun, in association with constitutively expressed c-Fos, increase levels of transcription factor activator-1 (AP-1), which is required for transcription of matrix metalloproteinases. Interestingly, the ERK MAP kinase pathway is reduced whereas the stress-activated MAP kinase pathway is increased in old, compared to young human skin *in vivo* (Chung et al., 2000). The transcription factor c-Jun is elevated two-fold in aged skin compared to young skin, suggesting c-Jun/AP1 may drive up-regulation of MMPs in aging skin (Chung et al., 2000). In addition, the c-Jun transcription factor negatively regulates type I procollagen (Fisher et al., 1998). Type I (α2) procollagen promoter activity is reduced by UV, and is further reduced by overexpression of wild type c-Jun. Overexpression of dominant negative mutant c-Jun completely abrogates UV inhibition of type I (α2) procollagen promoter activity (Fisher et al., 1998). It follows that UV inhibition of type I procollagen synthesis is mediated, at least in part, by UV-induced c-Jun. The coordinated regulation of collagen breakdown and collagen synthesis involves the transcription factor c-Jun. These data suggest that elevated c-Jun/AP1 activity in aged skin contributes to reduced collagen synthesis and increased collagen degradation in human skin connective tissue aging.

Altered Expression of CCN Family Proteins in Dermal Fibroblasts and Skin Connective Tissue Aging

Emerging evidence indicates that members of the CCN (CYR61/ CTGF/NOV) family proteins contribute to the altered collagen homeostasis in naturally-aged and photodamaged skin (Quan et al., 2002a; Quan et al., 2006; Quan et al., 2011a; Quan et al., 2012a; Quan et al., 2010b). Member of the CCN family at present comprises six distinct members: cysteine-rich protein 61 (CCN1), connective tissue growth factor (CCN2), nephroblastoma overexpressed (CCN3), Wnt-inducted secreted protein-1 (CCN4), Wnt-inducted secreted protein-2 (CCN5), and Wnt-inducted secreted protein-3 (CCN6) (Leask and Abraham, 2006; Perbal, 2004; Perbal et al., 2003). The CCN family proteins are secreted, ECM-associated heparin-binding proteins, and exhibit diverse cellular functions such as regulation of cell adhesion, proliferation, migration, chemotaxis, apoptosis, motility, and ECM remodeling in wound healing (Chen and Lau, 2009; Lau and Lam, 1999). It has been reported that CCN2 plays an important role in regulating procollagen synthesis (Brigstock, 1999; Duncan et al., 1999; Grotendorst, 1997; Gupta et al., 2000). CCN2 is rapidly induced by TGF-β, and appears to function as a downstream mediator of TGF-β actions (Quan et al., 2010b). CCN2 is markedly elevated in numerous fibrotic human disorders that involve skin, lungs, and kidneys (Abraham et al., 2000; Bradham et al., 1991; Grotendorst et al., 1996; Shi-Wen et al., 2008). CCN2, in conjunction with TGF-β, is believed to stimulate excessive deposition of collagen, the hallmark of fibrotic diseases. These findings have generated interest in development of new drugs that interfere with CCN2 function to combat fibrotic disorders. Although the role of CCN2 in the pathophysiology of fibrotic diseases has received considerable attention; the physiological role of CCN2 in the regulation of collagen expression in normal human skin has been only recently studied. Initially, CCN2 is detected in fibrotic skin diseases, such as scleroderma, but not in normal skin or cultured normal human dermal fibroblasts. However, CCN2 is later found to be readily detectable and constitutively expressed in normal human skin *in vivo* and normal human dermal fibroblasts (Quan et al., 2002a). In human dermal fibroblasts, CCN2 functions as a downstream mediator of TGF-β/Smad pathway, and is necessary for optimal TGF-β-dependent regulation of type I procollagen synthesis (Quan et al., 2010b). Importantly, CCN2 is predominantly expressed in dermal cells human, and is significantly reduced in fibroblasts in naturally-aged human skin *in vivo* (Quan et al., 2010b). Furthermore, acute UV irradiation reduces CCN2 expression in human skin *in vivo* and dermal fibroblast in culture (Quan et al., 2002a), suggesting that reduced CCN2 likely contributes to UV-induced photoaging in human skin. These data suggest that reduced CCN2 levels likely contribute to aberrant collagen homeostasis observed in naturally-aged and photodamaged human skin. Alterations of CCN2 expression are similar to those observed for TβRII, as described above. This similarity is consistent with TGF-β signaling being the major regulator of CCN2 expression.

In addition to CCN2, cysteine-rich protein 61 (CCN1), another member of the CCN family, functions as a novel mediator of collagen homeostasis (Chen et al., 2001; Quan et al., 2011a; Quan et al., 2012a). Like CCN2, CCN1 is predominantly expressed in dermal cells human (Quan et al., 2002a; Quan et al., 2006). However, unlike CCN2, CCN1 is significantly elevated in fibroblasts in naturally-aged and photoaged human skin *in vivo* (Quan et al., 2006; Quan et al., 2011a; Quan et al., 2010a).

In cultured human dermal fibroblasts, elevated expression of CCN1 substantially reduces type I procollagen, and concurrently increases MMP-1 (Quan et al., 2006). Interestingly, increased expression of CCN1 causes down-regulation of TβRII mRNA and protein levels, thereby negatively regulating the activity of TGF-β, in human dermal fibroblasts. Furthermore, elevated CCN1 induces transcription factor AP-1, which functions to stimulate MMP-1 expression. As replicative senescence is a form of cellular aging, CCN1 mRNA and protein levels are significantly elevated in replicative senescent dermal fibroblasts(Quan et al., 2012a).

Replicative senescent dermal fibroblasts also express significantly reduced levels of type I procollagen and increased levels of MMP-1, as observed in aged human skin *in vivo*. Knockdown of elevated CCN1 in senescent dermal fibroblasts partially normalizes both type I procollagen and MMP-1 expression. These data further support a key role of CCN1 in regulation of collagen homeostasis, and elevated CCN1 functions as an important mediator of collagen loss, which is observed in aged human skin.

CCN1 is also significantly elevated in acutely UV-irradiated human skin *in vivo*, and UV-irradiated human dermal fibroblasts (Quan et al., 2010a). Knockdown of CNN1 significantly attenuates UV irradiation-induced inhibition of type I procollagen and up-regulation of MMP-1, indicating CCN1 functions as a key mediator of UV-induced aberrant collagen homeostasis. Determination of CCN1 mRNA and protein indicates that the primary mechanism of CCN1 induction by UV irradiation is transcriptional. Analysis of CCN1 proximal promoter shows that a sequence conforming to the consensus binding site for transcription factor AP-1 is required for promoter activity. UV irradiation increases the binding of AP-1-family members c-Jun and c-Fos to CCN1 AP-1 binding site. Furthermore, functional blockade of c-Jun or knockdown of c-Jun significantly reduces the UV irradiation-induced activation of CCN1 promoter and CCN1 gene expression. These data show that CCN1 is transcriptionally regulated by UV irradiation through transcription factor AP-1, and mediates altered collagen homeostasis that occurs in response to UV irradiation in human dermal fibroblasts.

CCN1 contributes to human skin connective tissue photoaging, at least in part, through up-regulation of interleukin 1β (IL-1β) (Chen et al., 2001), (unpublished data). Both CCN1 and IL-1β are strikingly induced by acute UV irradiation in the dermis of human skin *in vivo*, and constitutively elevated in the dermis of chronic photoaged human skin. CCN1 markedly induces IL-1β, which in turn mediates CCN1-induced aberrant collagen homeostasis. Furthermore, knockdown of CCN1 significantly reduces UV-induced IL-1β, and therefore partially prevents UV-induced aberrant collagen homeostasis. These data demonstrate that induction of CCN1 by UV irradiation is responsible for up-regulation of IL-1β, which mediates UV-induced aberrant collagen homeostasis. Elevated expression of CCN1 is likely an important mediator of inflammation and connective tissue damage in photoaged human skin. Thus, elevated expression of CCN1 in human dermal fibroblasts acts through multiple pathways to reduce collagen production and increase MMP-1-mediated collagen fragmentation in naturally-aged and photoaged human skin *in vivo*.

Vitamin A and its metabolites have been shown to improve naturally-aged and photoaged skin by promoting deposition of new collagen and preventing its degradation (Kafi et al., 2007; Varani et al., 1994; Varani et al., 2000; Wang et al., 1999). In skin equivalent cultures, all-*trans* retinoic acid (RA), the major bioactive form of vitamin A in skin, significantly increases type I procollagen, and reduces MMP-1 (Quan et al., 2011b). Addition of recombinant human CCN1 to skin equivalent cultures, significantly reduces type I procollagen and increased MMP-1 (Quan et al., 2011b). Importantly, all-*trans* retinoic acid significantly reduces CCN1 expression in skin equivalent cultures (Quan et al., 2011b). Topical treatment with retinol (vitamin A, 0.4%) for seven days significantly reduces CCN1 and increase CCN2 (unpublished data) gene expression in both naturally-aged and photoaged human skin *in vivo*, compared to vehicle-treated skin (Quan et al., 2011b). These data indicate that the mechanism by which retinoids improve aged skin, through increased collagen production, involves down-regulation CCN1and up-regulation of CCN2.

CCN1 is not only induced by oxidative stress, but also increases intracellular levels of reactive oxygen species (ROS) (Jun and Lau, 2010; Quan et al., 2011b), and thereby accelerates the aging process (Quan et al., 2011a; Quan et al., 2012a). Elevated CCN1 in human dermis alters expression of numerous secreted proteins, which together have deleterious effects on dermal microenvironment, and thus impair the structural integrity and function of skin. CCN1-induced alteration of dermal microenvironment referred as "Age-Associated Secretory Phenotype (AASP)"(Quan et al., 2011a). AASP includes: 1) reduced expression skin ECM components, such as collagens related to dermal thinning; 2) increased expression of collagen-degrading enzymes, such as multiple MMPs related to fragmentation of dermal ECM; and 3) increased expression of pro-inflammatory cytokines, such as multiple interleukins related to oxidative stress tissue microenvironment. Importantly, CCN1-induced AASP is observed in naturally-aged and photoaged human dermis *in vivo*, and logically could account for many of the characteristic features of human skin. With natural aging and photoaging, progressive development of CCN1-induced AASP may drive loss of tissue structure and function, which are characteristic features of aging in all tissues including skin.

Emerging evidence indicates that the CCN proteins may represent a new class of modulators of inflammation (Jun and Lau, 2010; Kular et al., 2011). Consistent with this notion, CCN1 activates a proinflammatory genetic program in marine macrophages (Bai et al., 2010; Chen et al., 2001).

Furthermore, evidence indicates a potential role of CCN1 in chronic inflammatory diseases such as atherosclerosis, rheumatoid arthritis, inflammatory kidney diseases and neuroinflammatory diseases (Kular et al., 2011). There is strong association of aging with chronic low grade inflammatory activity which may progress to long term tissue damage and systemic chronic inflammation (Buckley, 2011; Douglas et al., 2002). Accumulating evidence supports a new concept called "inflammaging", which suggests that low grade chronic elevation of proinflammatory mediators can be a driving force for the aging progress (Franceschi, 2007; Franceschi et al., 2007; Goto, 2008; Lencel and Magne, 2011). Central to this concept is that healthy aging is not an inflammatory disease, but rather sub-clinical inflammation contributes to the gradual decline of organ function, which occurs during the aging process. For example, IL-6, which is an AASP protein, is increased in the aged and has been suggested to be a marker of health status in the elderly (Buckley, 2011; Franceschi, 2007; Maggio et al., 2006). Interestingly, IL-6 is elevated in aged skin and induced by CCN1 in human skin dermal fibroblasts (Quan et al., 2011a).

Although aged skin does not display overt inflammation, the functional role of subtle constitutive increases of cytokine levels in human skin connective tissue aging has not been investigated. It is conceivable that low level chronic elevation of IL-1β and IL-6 may contribute to sustained up-regulation of MMPs and collagen loss since these cytokines have been show to negatively regulate collagen homeostasis (Bauge et al., 2007; Franceschi et al., 2007).

Therefore, the possibility that CCN1-indicible cytokines contribute to human skin connective tissue aging deserves further investigation. It has been shown that CCN proteins functions primarily through interactions with integrins in a cell-type, function-specific manner (Chen and Lau, 2009; Katsube et al., 2009; Lau and Lam, 1999). CCN1 has been reported to interact with integrins to increase intracellular levels of ROS, which function as important effectors of CCN1actions (Jun and Lau, 2010).

The ability of CCN1 to induce AASP in dermal fibroblasts may be mediated by integrin-coupled generation of ROS. It is well-documented that the integrin pathway and ROS lead to stimulation of MAPK and NF-κb signaling, which up-regulate MMPs and proinflammatory cytokines, including IL-1β and IL-6 (Jun and Lau, 2010; Leask and Abraham, 2006). Therefore, it is conceivable that elevated CCN1 up-regulates MMPs and cytokines through activation of integrin and/or ROS-mediated activation of MAKP and NFκ-b signaling in aged human skin dermal fibroblasts.

It is also noteworthy that elevated IL-1β and IL-6 drives not only up-regulation of MMPs but also down-regulation of collagen synthesis, since these cytokines have been shown to negatively regulate collagen expression (Bauge et al., 2007; Franceschi et al., 2007).

Aged Dermal ECM Microenvironment and Age-Related Skin Diseases

As the US population ages, age-associated morbidity and diseases impose significant financial burdens on our nation's healthcare system (Gehrs et al., 2006; Hartman et al., 2008; Johnson, 1989). Indeed, healthcare cost per capita for the elderly population is over three times higher than that of a working-aged person. There is strong association of aging with increased risk of age-related common diseases, such as cancer, diabetes, atherosclerosis, cataracts, metabolic syndrome, sarcopenia osteoporosis, and decline in immune system function (Bourla and Young, 2006; Campisi, 2008; Kaufman, 2009; Kudravi and Reed, 2000). Skin aging is also associated with numerous deleterious alterations, such as impaired wound healing, increased frailty, greater risk of cancer development, and increased susceptibility to contact dermatitis (Fisher et al., 1997a; Holt et al., 1992; Kudravi and Reed, 2000; Zhao et al., 2010). The recent survey identified that skin cancer is the most common type of cancer in the United States, with more 3.5 million skin cancers in over two million people are diagnosed annually (American Cancer Society.; Rogers et al., 2010). Although mortality from skin cancer is low, skin cancer often entails significant medical costs, loss of function, and disfigurement. The annual cost of treatment in the US is estimated to be $1 billion. These findings are directly relevant to public health care in the largest sense, yet knowledge regarding the mechanisms by which solar UV radiation damages skin is far from complete.

As cancer is an age-related disease, one fundamental but critical question is that why the cancers more frequently occur in elderly but not in young population. For example, one in five Americans will develop skin cancer in the course of a lifetime, and 40-50% of Americans who lives to age 65 will have skin cancer at least once (Gloster and Brodland, 1996; National Cancer Institute (US), 2005). As development of skin cancer is associated with chronic solar exposure and age, it is possible that one of the major effects of chronic UV exposure on skin, alteration in the dermal ECM, may contribute to age-related skin carcinogenesis. Collagen degradation products are known to promote migration of such cells as vascular smooth muscle cells, wound healing keratinocytes and the immortal keratinocyte cell line, HaCaT (Carmeliet, 2000; Holt et al., 1992). Complex alterations of the composition of the ECM are associated with invasive melanoma (Aikawa et al., 2006; Grether-Beck and Krutmann, 2000; Krutmann, 2000). It is, thus, possible that UV alterations in the dermal collagen matrix in photodamaged skin may provide the microenvironment for the development of various invasive skin cancers. It has been suggested that alterations in dermal ECM/stroma that occur during skin aging process provide a microenvironment that supports the development of skin cancers (Fisher et al., 2009; Valera et al., 2000). Emerging evidence indicates that young tissue microenvironment (healthy architecture and normal tissue homeostasis) provides tumor-suppressive signals (Bissell and Hines, 2011; Campisi, 2008). However, as tissue homeostasis becomes disrupted with aging, such as AASP-associated dermal microenvironment in elderly, the tissue microenvironment can become a tumor promoter (Bissell and Hines, 2011). Indeed, recent publications indicate that alterations in the stromal microenvironment, such as loss of dermal fibroblast Notch (Hu et al., 2012) or TGF-β signaling (Bhowmick et al., 2004; Cheng et al., 2007), can induce epithelial tumorigenesis. Furthermore, elevated WNT (Sun et al., 2012) and HGF (Straussman et al., 2012) signaling in stromal fibroblasts can promote cancer drug resistance and control drug sensitivity. Although the influence of the stromal microenvironment on tumor progression is well recognized, little is known about the role of age-related aberrant dermal ECM microenvironment in skin cancer development.

Alteration of dermal ECM microenvironment such as collagen loss during aging increases fragility and impairs wound healing. Impaired wound healing is common in the elderly and presents a significant clinical and economic problem. Fibroblast dysfunction is a key factor in the non-healing of chronic wounds in the elderly (Mendez et al., 1998; Wall et al., 2008).

Dermal fibroblasts proliferation is reduced in response to wound in aged animals (Reed et al., 2001). Additionally, dermal collagen synthesis, angiogenesis, and epidermal reepithelialization all exhibit an age-related delay in response to wound in aged animals (Swift et al., 1999). There is a general decrease in the number and size of dermal fibroblasts with age, the primary cells that are responsible for collagen production and turnover in skin. In skin, dermal fibroblasts interact with and attach to three dimensional (3D) collagen fibrils, thereby maintain normal cell shape/mechanical tension for function (Fisher et al., 2008).

Both in aged human skin *in vivo* and *in vitro* 3D collagen matrix, dermal fibroblasts are unable to maintain normal cell shape/mechanical tension in damaged collagen microenvironment, and therefore lost their functions in collagen homeostasis; produce more MMPs and less collagen, that are central event of the skin connective tissue aging (Fisher et al., 2009; Fisher et al., 2008). It has been shown that dermal fibroblasts cultured from individuals of different age are largely indistinguishable in collagen homeostasis *in vitro* monolayer culture (Fisher et al., 2009).

However, in 3D culture, fragmented collagen alters dermal fibroblast function to shift the balance to produce more MMPs and less collagen in aged skin (Fisher et al., 2009). ECM microenvironment of the dermis is largely responsible for the morphological and functional differences that observed between fibroblasts in young versus aged human skin *in vivo* (Fisher et al., 2009; Fisher et al., 2008; Quan et al., 2012b). Therefore, it is conceivable that alteration of dermal collagen network due to loss of dermal collagen structural and mechanical properties in aged skin is unable to support normal mechanical tension within fibroblasts.

In contrast, age-related reduced production of collagen is largely reversed by enhancing structural support of the ECM (Quan et al., 2012b; Wang et al., 2007). Injection of dermal filler, cross-linked hyaluronic acid, into the aged skin stimulates fibroblasts to produce type I collagen. This stimulation is associated with localized increased of mechanical forces, indicated by fibroblast elongation/spreading, and mediated by up-regulation of TβRII and CCN2.

Interestingly, enhanced mechanical support of the collagen ECM also stimulates fibroblast proliferation, expands vasculature, and increases epidermal thickness. Injection of filler into dermal equivalent cultures causes elongation of fibroblasts, coupled with type I collagen synthesis, which is dependent on the TGF-β signaling pathway.

Thus, fibroblasts in aged human skin retain their capacity for functional activation, which is restored by enhancing structural support of the ECM.

As the actual mechanical properties of the skin reside mainly in the collagen, a wealth of evidence indicates that mechanical properties of tissue microenvironment regulate a variety of cellular processes including signal transduction, gene expression, and tissue homeostasis (Fisher et al., 2009; Ingber, 1997a; Spencer et al., 2007; Underhill et al., 2012; Varani et al., 2004).

Alterations of the dermal ECM mechanical microenvironment in aged skin have significant consequences on regulation of collagen homeostasis by fibroblasts (Fisher et al., 2009; Quan et al., 2012b). Currently, it is not clear that how the cells sense and transduce signals in response to changes of surrounding microenvironment, such as damaged collagen network in aged skin. Fragmented collagen is unable to support normal cell shape and mechanical tension within fibroblasts, and this loss of cell shape and mechanical tension is closely associated with increased c-Jun and transcription factor AP-1(Fisher et al., 2009). As transcription factor c-Jun/AP-1 functions as a major driving force for multiple MMPs and potent negative regulator of type I procollagen expression (Fisher et al., 1996b; Fisher et al., 1998; Quan et al., 2010a), it is conceivable that elevated AP-1 activity significantly contributes to the multiple elevated MMPs and loss of type I collagen expression in aged skin.

Additionally, alterations of dermal collagen network in aged skin may also impair mechanical properties of the adhesion substrate, ECM-integrin mechanical sensing, and subsequent integrin signaling events associated with aberrant collagen homeostasis in aged skin.

A wealth of information indicates that the interaction of cells with their surrounding ECM environment is critical for cellular function (Ingber, 1997a; Ingber, 1997b; Lambert et al., 2001).

In human skin, dermal fibroblasts physically interact with and attach to the three-dimensional collagen network through a class of cell membrane receptor, integrins (Boudreau and Bissell, 1998; Giancotti and Ruoslahti, 1999; Miranti and Brugge, 2002). Integrin receptors form structural and functional linkages between the ECM and intracellular cytoskeletal linker proteins. Integrins interact with the actin/myosin-based cytoskeleton to form cellular machinery that is capable of physically contracting the pliable ECM. This contraction operates against the tensile properties of the ECM to generate mechanical forces which define cell morphology.

Alterations of collagen network in aged dermis impair mechanical properties of the adhesion substrate, ECM-integrin mechanical sensing, and subsequent integrin signaling events.

Damaged collagen network in aged skin is unable to support normal mechanical tension within fibroblasts, and this loss of mechanical tension is closely associated with altered dermal fibroblast function in collagen homeostasis.

Conclusion and Outlook

Aberrant dermal collagen homeostasis, which is caused by impaired dermal fibroblast function, is primary responsible for the human skin connective tissue aging.

Figure 2 depicts a model for how aged dermal fibroblasts contribute to human skin connective tissue aging and age-related skin diseases. Dermal fibroblast functions in aged skin are significantly impaired in collagen homeostasis as evidence of impaired TGF-β signaling, the major regulator of collagen production, and altered expression of CCN family proteins, the novel mediators of collagen homeostasis. This age-dependent loss of dermal fibroblast collagen homeostasis causes dermal collagen loss and accumulation of damaged collagen, which is characteristic feature of skin connective tissue aging.

Loss of dermal collagen in aged skin alters dermal ECM microenvironment by reducing structural integrity and mechanical properties. Reduced structural integrity and mechanical properties create a tissue environment that causes age-related skin diseases, such as increased fragility and poor wound healing, and promotes skin cancer. These structural and mechanical alterations of dermis likely also impact epidermal function, contributing to epidermal thinning, which is a prominent feature of aged skin.

On the other hand, alteration of dermal collagen structural and mechanical environment has a significant impact on dermal fibroblasts function by reducing attachment to collagen fibers, reduced spreading, and reduced mechanical tension.

Therefore, this working model reveals an inherent self-perpetuating positive feedback nature between impaired dermal fibroblast function and alteration of dermal collagen structural and mechanical environment. Aged dermal fibroblast weakens the structural integrity and mechanical properties of the dermal ECM.

The weakened dermal ECM microenvironment provides less resistance to the mechanical forces exerted on it by dermal fibroblasts, thereby reducing the mechanical tension in the fibroblasts. This reduced mechanical tension in the fibroblasts results in further impairing dermal fibroblast function in collagen homeostasis and weakening the structural integrity and mechanical properties of the dermal ECM. The essential positive feedback between impaired dermal fibroblast function and loss of dermal collagen network is consistent with the biology of aging, which is represented by gradual and progressive loss of homeostatic control within tissues.

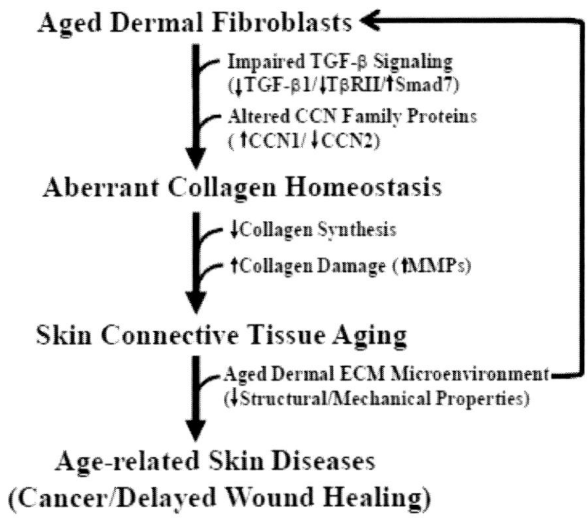

Figure 2. A model for the aged dermal fibroblasts contributes to human skin connective tissue aging and age-related skin diseases (see Conclusion and outlook in details).

Acknowledgments

This study is supported by the National Institutes of Health (Bethesda, MD) Grants: ES014697 and ES014697 30S1 to Taihao Quan, AG019364 to Gary J. Fisher and Taihao Quan, AG031452 and AG025186 to Gary J. Fisher. The author would like to thank Drs Gary J. Fisher and John J. Voorhees for their help and support. The author also thanks Drs Zhaping Qin, Tianyuan He, Yuan Shao, and Trupta Purohit for technical assistance.

Rererences

(2010) Sun Protection. *Cancer Trends Progress Report* – 2009/2010 Update. National Cancer Institute. http://progressreport.cancer.gov/doc_detail.asp? pid=1anddid=2007andchid=71 andcoid=711andmid. Accessed.

Abraham, D. J., Shiwen, X., Black, C. M., Sa, S., Xu, Y., Leask, A. (2000) Tumor necrosis factor alpha suppresses the induction of connective tissue growth factor by transforming growth factor-beta in normal and scleroderma fibroblasts. *J. Biol. Chem.* 275:15220-5.

Afrakhte, M., Moren, A., Jossan, S., Itoh, S., Sampath, K., Westermark, B., *et al.* (1998) Induction of inhibitory Smad6 and Smad7 mRNA by TGF-beta family members. *Biochem. Biophys. Res. Commun.* 249:505-11.

Aikawa, T., Gunn, J., Spong, S. M., Klaus, S. J., Korc, M. (2006) Connective tissue growth factor-specific antibody attenuates tumor growth, metastasis, and angiogenesis in an orthotopic mouse model of pancreatic cancer. *Mol. Cancer Ther.* 5:1108-16.

Bai, T., Chen, C. C., Lau, L. F. (2010) Matricellular protein CCN1 activates a proinflammatory genetic program in murine macrophages. *J. Immunol.* 184:3223-32.

Bauge, C., Legendre, F., Leclercq, S., Elissalde, J. M., Pujol, J. P., Galera, P., *et al.* (2007) Interleukin-1beta impairment of transforming growth factor beta1 signaling by down-regulation of transforming growth factor beta receptor type II and up-regulation of Smad7 in human articular chondrocytes. *Arthritis Rheum.* 56:3020-32.

Bernstein, E. F., Chen, Y. Q., Kopp, J. B., Fisher, L., Brown, D. B., Hahn, P. J., *et al.* (1996) Long-term sun exposure alters the collagen of the papillary dermis. Comparison of sun-protected and photoaged skin by northern analysis, immunohistochemical staining, and confocal laser scanning microscopy. *J. Am. Acad. Dermatol.* 34:209-18.

Bhowmick NA, Chytil A, Plieth D, Gorska AE, Dumont N, Shappell S, et al. (2004) TGF-beta signaling in fibroblasts modulates the oncogenic potential of adjacent epithelia. Science 303:848-51.

Biesalski, H. K., Berneburg, M., Grune, T., Kerscher, M., Krutmann, J., Raab, W., *et al.* (2003) Hohenheimer Consensus Talk. Oxidative and premature skin ageing. *Exp. Dermatol.* 12 Suppl. 3:3-15.

Birkedal-Hansen, H., Moore, W. G., Bodden, M. K., Windsor, L. J., Birkedal-Hansen, B., DeCarlo, A., *et al.* (1993) Matrix metalloproteinases: a review. *Crit. Rev. Oral Biol. Med.* 4:197-250.

Bissell, M. J., Hines, W. C. (2011) Why don't we get more cancer? A proposed role of the microenvironment in restraining cancer progression. *Nat. Med.* 17:320-9.

Bolognia, J., Jorizzo, J. L., Rapini, R. P. (2008) *Dermatology.* (Mosby/ Elsevier: [St. Louis, Mo.].

Boudreau, N., Bissell, M. (1998) Extracellular matrix signaling: integration of form and function in normal and malignant cells. *Curr. Opin. Cell. Biol.* 10:640-6.

Bourla, D. H., Young, T. A. (2006) Age-related macular degeneration: a practical approach to a challenging disease. *J. Am. Geriatr. Soc.* 54:1130-5.

Bradham, D. M., Igarashi, A., Potter, R. L., Grotendorst, G. R. (1991) Connective tissue growth factor: a cysteine-rich mitogen secreted by human vascular endothelial cells is related to the SRC-induced immediate early gene product CEF-10. *J. Cell. Biol.* 114:1285-94.

Brennan, M., Bhatti, H., Nerusu, K. C., Bhagavathula, N., Kang, S., Fisher, G. J., *et al.* (2003) Matrix metalloproteinase-1 is the major collagenolytic enzyme responsible for collagen damage in UV-irradiated human skin. *Photochem. Photobiol.* 78:43-8.

Brenneisen, P., Sies, H., Scharffetter-Kochanek, K. (2002) Ultraviolet-B irradiation and matrix metalloproteinases: from induction via signaling to initial events. *Ann. N Y Acad. Sci.* 973:31-43.

Brigstock, D. R. (1999) The connective tissue growth factor/cysteine-rich 61/nephroblastoma overexpressed (CCN) family. *Endocr. Rev.* 20:189-206.

Brodin, G., Ahgren, A., ten Dijke, P., Heldin, C. H., Heuchel, R. (2000) Efficient TGF-beta induction of the Smad7 gene requires cooperation between AP-1, Sp1, and Smad proteins on the mouse Smad7 promoter. *J. Biol. Chem.* 275:29023-30.

Buckley, C. D. (2011) Why does chronic inflammation persist: An unexpected role for fibroblasts. *Immunol. Lett.* 138:12-4.

Campisi, J. (2008) Aging and cancer cell biology, 2008. *Aging Cell* 7:281-4.

Carmeliet, P. (2000) Mechanisms of angiogenesis and arteriogenesis. *Nat. Med.* 6:389-95.

Chen, C. C., Lau, L. F. (2009) Functions and mechanisms of action of CCN matricellular proteins. *Int. J. Biochem. Cell Biol.* 41:771-83.

Chen, C. C., Mo, F. E., Lau, L. F. (2001) The angiogenic factor Cyr61 activates a genetic program for wound healing in human skin fibroblasts. *J. Biol. Chem.* 276:47329-37.

Cheng N, Chytil A, Shyr Y, Joly A, Moses HL (2007) Enhanced hepatocyte growth factor signaling by type II transforming growth factor-beta receptor knockout fibroblasts promotes mammary tumorigenesis. Cancer Res 67:4869-77.

Chung, J. H., Kang, S., Varani, J., Lin, J., Fisher, G. J., Voorhees, J. J. (2000) Decreased extracellular-signal-regulated kinase and increased stress-activated MAP kinase activities in aged human skin in vivo. *J. Invest. Dermatol.* 115:177-82.

Chuong, C. M., Nickoloff, B. J., Elias, P. M., Goldsmith, L. A., Macher, E., Maderson, P. A., *et al.* (2002) What is the 'true' function of skin? *Exp. Dermatol.* 11:159-87.

Cuchacovich, T. M., Pacheco, P., Merino, G., Gallardo, P., Gatica, H., Valenzuela, H., *et al.* (2000) [Clinical features and response to systemic treatment of primary and secondary episcleritis and scleritis resistant to local treatment]. *Rev. Med. Chil.* 128:1205-14.

Di Lullo, G. A., Sweeney, S. M., Korkko, J., Ala-Kokko, L., San Antonio, J. D. (2002) Mapping the ligand-binding sites and disease-associated mutations on the most abundant protein in the human, type I collagen. *J. Biol. Chem.* 277:4223-31.

Dong, C., Zhu, S., Wang, T., Yoon, W., Li, Z., Alvarez, R. J., *et al.* (2002) Deficient Smad7 expression: a putative molecular defect in scleroderma. *Proc. Natl. Acad. Sci. US* 99:3908-13.

Douglas, M. R., Morrison, K. E., Salmon, M., Buckley, C. D. (2002) Why does inflammation persist: a dominant role for the stromal microenvironment? *Expert Rev. Mol. Med.* 4:1-18.

Duncan, M. R., Frazier, K. S., Abramson, S., Williams, S., Klapper, H., Huang, X., *et al.* (1999) Connective tissue growth factor mediates transforming growth factor beta-induced collagen synthesis: down-regulation by cAMP. *FASEB J.* 13:1774-86.

Egeblad, M., Werb, Z. (2002) New functions for the matrix metalloproteinases in cancer progression. *Nat. Rev. Cancer* 2:161-74.

Farage, M. A., Miller, K. W., Elsner, P., Maibach, H. I. (2008) Functional and physiological characteristics of the aging skin. *Aging Clin. Exp. Res.* 20: 195-200.

Farage, M. A., Miller, K. W., Maibach, H. I., SpringerLink (Online service) (2010) *Textbook of Aging Skin.* (Springer-Verlag Berlin Heidelberg: Berlin, Heidelberg.

Fiocchi, C. (2001) TGF-beta/Smad signaling defects in inflammatory bowel disease: mechanisms and possible novel therapies for chronic inflammation. *J. Clin. Invest.* 108:523-6.

Fisher, G. J., Datta, S., Wang, Z., Li, X. Y., Quan, T., Chung, J. H., *et al.* (2000) c-Jun-dependent inhibition of cutaneous procollagen transcription following ultraviolet irradiation is reversed by all-trans retinoic acid. *J. Clin. Invest.* 106:663-70.

Fisher, G. J., Datta, S. C., Talwar, H. S., Wang, Z. Q., Varani, J., Kang, S., *et al.* (1996a) Molecular basis of sun-induced premature skin ageing and retinoid antagonism. *Nature* 379:335-9.

Fisher, G. J., Datta, S. C., Talwar, H. S., Wang, Z. Q., Varani, J., Kang, S., *et al.* (1996b) Molecular basis of sun-induced premature skin ageing and retinoid antagonism. *Nature* 379:335-9.

Fisher, G. J., Kang, S., Varani, J., Bata-Csorgo, Z., Wan, Y., Datta, S., *et al.* (2002) Mechanisms of photoaging and chronological skin aging. *Arch. Dermatol.* 138:1462-70.

Fisher, G. J., Quan, T., Purohit, T., Shao, Y., Cho, M. K., He, T., *et al.* (2009) Collagen fragmentation promotes oxidative stress and elevates matrix metalloproteinase-1 in fibroblasts in aged human skin. *Am. J. Pathol.* 174: 101-14.

Fisher, G. J., Talwar, H. S., Lin, J., Lin, P., McPhillips, F., Wang, Z., *et al.* (1998) Retinoic acid inhibits induction of c-Jun protein by ultraviolet radiation that occurs subsequent to activation of mitogen-activated protein kinase pathways in human skin in vivo. *J. Clin. Invest.* 101:1432-40.

Fisher, G. J., Varani, J., Voorhees, J. J. (2008) Looking older: fibroblast collapse and therapeutic implications. *Arch. Dermatol.* 144:666-72.

Fisher, G. J., Wang, Z. Q., Datta, S. C., Varani, J., Kang, S., Voorhees, J. J. (1997a) Pathophysiology of premature skin aging induced by ultraviolet light. *N Engl. J. Med.* 337:1419-28.

Fisher, G. J., Wang, Z. Q., Datta, S. C., Varani, J., Kang, S., Voorhees, J. J. (1997b) Pathophysiology of premature skin aging induced by ultraviolet light. *New Eng. J. Med.* 337:1419-28.

Fligiel, S. E., Varani, J., Datta, S. C., Kang, S., Fisher, G. J., Voorhees, J. J. (2003) Collagen degradation in aged/photodamaged skin in vivo and after exposure to matrix metalloproteinase-1 in vitro. *J. Invest. Dermatol.* 120: 842-8.

Franceschi, C. (2007) Inflammaging as a major characteristic of old people: can it be prevented or cured? *Nutr. Rev.* 65:S173-6.

Franceschi, C., Capri, M., Monti, D., Giunta, S., Olivieri, F., Sevini, F., *et al.* (2007) Inflammaging and anti-inflammaging: a systemic perspective on aging and longevity emerged from studies in humans. *Mech. Ageing Dev.* 128:92-105.

Fu, X., Parks, W. C., Heinecke, J. W. (2008) Activation and silencing of matrix metalloproteinases. *Semin. Cell Dev. Biol.* 19:2-13.

Gabbiani, G. (2003) The myofibroblast in wound healing and fibrocontractive diseases. *Journal of Pathology* 200:500-3.

Gehrs, K. M., Anderson, D. H., Johnson, L. V., Hageman, G. S. (2006) Age-related macular degeneration--emerging pathogenetic and therapeutic concepts. *Ann. Med.* 38:450-71.

Giancotti, F., Ruoslahti, E. (1999) Integrin Signaling. *Science* 285:1028-32.

Gloster, H. M., Jr., Brodland, D. G. (1996) The epidemiology of skin cancer. *Dermatol. Surg.* 22:217-26.

Gomez, D., Alonso, H., Yoshiji, H., Thorgeirsson, U. (1997) Tissue inhibitors of metalloproteinases: structure, regulation and biological functions. *Eur. J. Cell Biol.* 74:111-2.

Goto, M. (2008) Inflammaging (inflammation + aging): A driving force for human aging based on an evolutionarily antagonistic pleiotropy theory? *Biosci. Trends* 2:218-30.

Grether-Beck, S., Krutmann, J. (2000) Gene regulation by ultraviolet A radiation and singlet oxygen. *Methods Enzymol.* 319:280-90.

Grotendorst, G. R. (1997) Connective tissue growth factor: a mediator of TGF-beta action on fibroblasts. *Cytokine Growth Factor Rev.* 8:171-9.

Grotendorst, G. R., Okochi, H., Hayashi, N. (1996) A novel transforming growth factor beta response element controls the expression of the connective tissue growth factor gene. *Cell Growth Differ.* 7:469-80.

Gupta, S., Clarkson, M. R., Duggan, J., Brady, H. R. (2000) Connective tissue growth factor: potential role in glomerulosclerosis and tubulointerstitial fibrosis. *Kidney Int.* 58:1389-99.

Hall, M. C., Young, D. A., Waters, J. G., Rowan, A. D., Chantry, A., Edwards, D. R., *et al.* (2003) The comparative role of activator protein 1 and Smad factors in the regulation of

Timp-1 and MMP-1 gene expression by transforming growth factor-beta 1. *J. Biol. Chem.* 278:10304-13.

Hartman, M., Catlin, A., Lassman, D., Cylus, J., Heffler, S. (2008) US Health spending by age, selected years through 2004. *Health Aff. (Millwood)* 27:w1-w12.

Hayashi, H., Abdollah, S., Qiu, Y., Cai, J., Xu, Y. Y., Grinnell, B. W., *et al.* (1997) The MAD-related protein Smad7 associates with the TGFbeta receptor and functions as an antagonist of TGFbeta signaling. *Cell* 89: 1165-73.

He, W., Li, A. G., Wang, D., Han, S., Zheng, B., Goumans, M. J., *et al.* (2002) Overexpression of Smad7 results in severe pathological alterations in multiple epithelial tissues. *EMBO J.* 21:2580-90.

Hegedus, L., Cho, H., Xie, X., Eliceiri, G. L. (2008) Additional MDA-MB-231 breast cancer cell matrix metalloproteinases promote invasiveness. *J. Cell Physiol.* 216:480-5.

Holt, D. R., Kirk, S. J., Regan, M. C., Hurson, M., Lindblad, W. J., Barbul, A. (1992) Effect of age on wound healing in healthy human beings. *Surgery* 112:293-7; discussion 7-8.

Hu B, Castillo E, Harewood L, Ostano P, Reymond A, Dummer R, et al. (2012) Multifocal epithelial tumors and field cancerization from loss of mesenchymal CSL signaling. Cell 149:1207-20.

Hurskainen, R., Salmi, A., Paavonen, J., Teperi, J., Rutanen, E. (2000) Expression of sex steroid receptors and Ki-67 in the endometria of menorrhagic women: effects of intrauterine levonorgestrel. *Mol. Hum. Reprod.* 6:1013-8.

Ignotz, R.A., Massague, J. (1986) Transforming growth factor ß stimulates the expression of fibronectin and collagen and their incorporation into the extracellular matrix. *J. Biol. Chem.* 261:4337-45.

Ingber, D. E. (1997a) Integrins, tensegrity, and mechanotransduction. *Gravit. Space Biol. Bull.* 10:49-55.

Ingber, D. E. (1997b) Tensegrity: the architectural basis of cellular mechanotransduction. *Annu. Rev. Physiol.* 59:575-99.

Itoh, S., Landstrom, M., Hermansson, A., Itoh, F., Heldin, C. H., Heldin, N. E., *et al.* (1998) Transforming growth factor beta1 induces nuclear export of inhibitory Smad7. *J. Biol. Chem.* 273:29195-201.

Johnson, M. L. (1989) Aging of the United States population. The dermatologic implications. *Clin. Geriatr. Med.* 5:41-51.

Jun, J. I., Lau, L. F. (2010) The matricellular protein CCN1 induces fibroblast senescence and restricts fibrosis in cutaneous wound healing. *Nat. Cell Biol.* 12:676-85.

Kafi, R., Kwak, H. S., Schumacher, W. E., Cho, S., Hanft, V. N., Hamilton, T. A., *et al.* (2007) Improvement of naturally aged skin with vitamin A (retinol). *Arch. Dermatol.* 143:606-12.

Katsube, K., Sakamoto, K., Tamamura, Y., Yamaguchi, A. (2009) Role of CCN, a vertebrate specific gene family, in development. *Dev. Growth Differ.* 51:55-67.

Kaufman, S. R. (2009) Developments in age-related macular degeneration: Diagnosis and treatment. *Geriatrics* 64:16-9.

Kerkela, E., Saarialho-Kere, U. (2003) Matrix metalloproteinases in tumor progression: focus on basal and squamous cell skin cancer. *Exp. Dermatol.* 12:109-25.

Krutmann, J. (2000) Ultraviolet A radiation-induced biological effects in human skin: relevance for photoaging and photodermatosis. *J. Dermatol. Sci.* 23 Suppl. 1:S22-6.

Kudravi, S. A., Reed, M. J. (2000) Aging, cancer, and wound healing. *In Vivo* 14:83-92.

Kular, L., Pakradouni, J., Kitabgi, P., Laurent, M., Martinerie, C. (2011) The CCN family: a new class of inflammation modulators? *Biochimie* 93:377-88.

Lambert, C. A., Colige, A. C., Munaut, C., Lapiere, C. M., Nusgens, B. V. (2001) Distinct pathways in the over-expression of matrix metalloproteinases in human fibroblasts by relaxation of mechanical tension. *Matrix. Biol.* 20:397-408.

Lapiere, Ch M. (2005) Tadpole collagenase, the single parent of such a large family. *Biochimie* 87:243-7.

Lau, L. F., Lam, S. C. (1999) The CCN family of angiogenic regulators: the integrin connection. *Exp. Cell Res.* 248:44-57.

Leask, A., Abraham, D. J. (2006) All in the CCN family: essential matricellular signaling modulators emerge from the bunker. *J. Cell Sci.* 119: 4803-10.

Lencel, P., Magne, D. (2011) Inflammaging: the driving force in osteoporosis? *Med. Hypotheses* 76:317-21.

Maggio, M., Guralnik, J. M., Longo, D. L., Ferrucci, L. (2006) Interleukin-6 in aging and chronic disease: a magnificent pathway. *J. Gerontol. A Biol. Sci. Med. Sci.* 61:575-84.

Massague, J. (1990) The transforming growth factor-beta family. *Annu. Rev. Cell. Biol.* 6:597-641.

Massague, J. (2000) How cells read TGF-beta signals. *Nat. Rev. Mol. Cell. Biol.* 1:169-78.

Mauviel, A. (1993) Cytokine regulation of metalloproteinase gene expression. *J. Cell. Biochem.* 53:288-95.

Mendez, M. V., Stanley, A., Park, H. Y., Shon, K., Phillips, T., Menzoian, J. O. (1998) Fibroblasts cultured from venous ulcers display cellular characteristics of senescence. *J. Vasc. Surg.* 28:876-83.

Miranti, C., Brugge, J. (2002) Sensing the environment: a historical perspective on integrin signal transduction. *Nature Cell Biol.* 4:83-90.

Monteleone, G., Kumberova, A., Croft, N. M., McKenzie, C., Steer, H. W., MacDonald, T. T. (2001) Blocking Smad7 restores TGF-beta1 signaling in chronic inflammatory bowel disease. *J. Clin. Invest.* 108:601-9.

Nagarajan, R. P., Zhang, J., Li, W., Chen, Y. (1999) Regulation of Smad7 promoter by direct association with Smad3 and Smad4. *J. Biol. Chem.* 274:33412-8.

Nakao, A., Afrakhte, M., Moren, A., Nakayama, T., Christian, J. L., Heuchel, R., *et al.* (1997) Identification of Smad7, a TGFbeta-inducible antagonist of TGF-beta signalling. *Nature* 389:631-5.

Nakao, A., Okumura, K., Ogawa, H. (2002) Smad7: a new key player in TGF-beta-associated disease. *Trends Mol. Med.* 8:361-3.

National Cancer Institute (US) (2005) Cancer trends progress report. (National Cancer Institute, NIH, DHHS: Bethesda, MD.

Paalman, M. H., Cotton, R. G., Kazazian, H. H., Jr. (2000) VARIATION, DATABASES, and DISEASE: new directions for human mutation. *Hum. Mutat.* 16:97-8.

Page-McCaw, A., Ewald, A. J., Werb, Z. (2007) Matrix metalloproteinases and the regulation of tissue remodelling. *Nat. Rev. Mol. Cell Biol.* 8:221-33.

Patil, A. S., Sable, R. B., Kothari, R. M. (2011) An update on transforming growth factor-beta (TGF-beta): sources, types, functions and clinical applicability for cartilage/bone healing. *J. Cell. Physiol.* 226:3094-103.

Perbal, B. (2004) CCN proteins: multifunctional signalling regulators. *Lancet* 363:62-4.

Perbal, B., Brigstock, D. R., Lau, L. F. (2003) Report on the second international workshop on the CCN family of genes. *Mol. Pathol.* 56:80-5.

Perumal, S., Antipova, O., Orgel, J.P. (2008) Collagen fibril architecture, domain organization, and triple-helical conformation govern its proteolysis. *Proc. Natl. Acad. Sci. US* 105:2824-9.

Quan, T., He, T., Kang, S., Voorhees, J. J., Fisher, G. J. (2002a) Connective tissue growth factor: expression in human skin in vivo and inhibition by ultraviolet irradiation. *J. Invest. Dermatol.* 118:402-8.

Quan, T., He, T., Kang, S., Voorhees, J. J., Fisher, G. J. (2002b) Ultraviolet irradiation alters transforming growth factor beta/smad pathway in human skin in vivo. *J. Invest. Dermatol.* 119:499-506.

Quan, T., He, T., Kang, S., Voorhees, J. J., Fisher, G. J. (2004) Solar ultraviolet irradiation reduces collagen in photoaged human skin by blocking transforming growth factor-beta type II receptor/Smad signaling. *Am. J. Pathol.* 165:741-51.

Quan, T., He, T., Shao, Y., Lin, L., Kang, S., Voorhees, J. J., *et al.* (2006) Elevated cysteine-rich 61 mediates aberrant collagen homeostasis in chronologically aged and photoaged human skin. *Am. J. Pathol.* 169:482-90.

Quan, T., He, T., Voorhees, J. J., Fisher, G. J. (2001) Ultraviolet irradiation blocks cellular responses to transforming growth factor-beta by down-regulating its type-II receptor and inducing Smad7. *J. Biol. Chem.* 276: 26349-56.

Quan, T., He, T., Voorhees, J. J., Fisher, G. J. (2005) Ultraviolet irradiation induces Smad7 via induction of transcription factor AP-1 in human skin fibroblasts. *J. Biol. Chem.* 280:8079-85.

Quan, T., Qin, Z., Robichaud, P., Voorhees, J. J., Fisher, G. J. (2011a) CCN1 contributes to skin connective tissue aging by inducing age-associated secretory phenotype in human skin dermal fibroblasts. *J. Cell Commun. Signal.* 5:201-7.

Quan, T., Qin, Z., Shao, Y., Xu, Y., Voorhees, J. J., Fisher, G. J. (2011b) Retinoids suppress cysteine-rich protein 61 (CCN1), a negative regulator of collagen homeostasis, in skin equivalent cultures and aged human skin in vivo. *Exp. Dermatol.* 20:572-6.

Quan, T., Qin, Z., Voorhees, J. J., Fisher, G. J. (2012a) Cysteine-rich protein 61 (CCN1) mediates replicative senescence-associated aberrant collagen homeostasis in human skin fibroblasts. *J. Cell. Biochem.*

Quan, T., Qin, Z., Xia, W., Shao, Y., Voorhees, J. J., Fisher, G. J. (2009) Matrix-degrading metalloproteinases in photoaging. *J. Investig. Dermatol. Symp. Proc.* 14:20-4.

Quan, T., Qin, Z., Xu, Y., He, T., Kang, S., Voorhees, J. J., *et al.* (2010a) Ultraviolet irradiation induces CYR61/CCN1, a mediator of collagen homeostasis, through activation of transcription factor AP-1 in human skin fibroblasts. *J. Invest. Dermatol.* 130:1697-706.

Quan, T., Shao, Y., He, T., Voorhees, J. J., Fisher, G. J. (2010b) Reduced expression of connective tissue growth factor (CTGF/CCN2) mediates collagen loss in chronologically aged human skin. *J. Invest. Dermatol.* 130:415-24.

Quan, T., Wang, F., Shao, Y., Rittie, L., Xia, W., Orringer, J. S., *et al.* (2012b) Enhancing Structural Support of the Dermal Microenvironment Activates Fibroblasts, Endothelial Cells, and Keratinocytes in Aged Human Skin In Vivo. *J. Invest. Dermatol.*

Reed, M. J., Ferara, N. S., Vernon, R. B. (2001) Impaired migration, integrin function, and actin cytoskeletal organization in dermal fibroblasts from a subset of aged human donors. *Mech. Ageing Dev.* 122:1203-20.

Rittie, L., Fisher, G. J. (2002) UV-light-induced signal cascades and skin aging. *Ageing Res. Rev.* 1:705-20.

Rodriguez C, Rodriguez-Sinovas A, Martinez-Gonzalez J (2008) Lysyl oxidase as a potential therapeutic target. *Drug News Perspect* 21:218-24.

Rogers HW, Weinstock MA, Harris AR, Hinckley MR, Feldman SR, Fleischer AB, *et al.* (2010) Incidence estimate of nonmelanoma skin cancer in the United States, 2006. *Arch Dermatol* 146:283-7.

Shamis, Y., Hewitt, K. J., Carlson, M. W., Margvelashvilli, M., Dong, S., Kuo, C. K., *et al.* (2011) Fibroblasts derived from human embryonic stem cells direct development and repair of 3D human skin equivalents. *Stem Cell. Res. Ther.* 2:10.

Shi-Wen, X., Leask, A., Abraham, D. (2008) Regulation and function of connective tissue growth factor/CCN2 in tissue repair, scarring and fibrosis. *Cytokine Growth Factor Rev.* 19:133-44.

Spencer, V. A., Xu, R., Bissell, M. J. (2007) Extracellular matrix, nuclear and chromatin structure, and gene expression in normal tissues and malignant tumors: a work in progress. *Adv. Cancer Res.* 97:275-94.

Stamenkovic, I. (2003) Extracellular matrix remodelling: the role of matrix metalloproteinases. *J. Pathol.* 200:448-64.

Sternlicht, M. D., Werb, Z. (2001) How matrix metalloproteinases regulate cell behavior. *Annu. Rev. Cell. Dev. Biol.* 17:463-516.

Stopa, M., Anhuf, D., Terstegen, L., Gatsios, P., Gressner, A. M., Dooley, S. (2000) Participation of Smad2, Smad3, and Smad4 in transforming growth factor beta (TGF-beta)-induced activation of Smad7. THE TGF-beta response element of the promoter requires functional Smad binding element and E-box sequences for transcriptional regulation. *J. Biol. Chem.* 275:29308-17.

Straussman R, Morikawa T, Shee K, Barzily-Rokni M, Qian ZR, Du J, et al. (2012) Tumour micro-environment elicits innate resistance to RAF inhibitors through HGF secretion. Nature 487:500-4.

Sun Y, Campisi J, Higano C, Beer TM, Porter P, Coleman I, et al. (2012) Treatment-induced damage to the tumor microenvironment promotes prostate cancer therapy resistance through WNT16B. Nat Med.

Swift, M. E., Kleinman, H. K., DiPietro, L. A. (1999) Impaired wound repair and delayed angiogenesis in aged mice. *Lab. Invest.* 79:1479-87.

Szpak, P. (2011) Fish bone chemistry and ultrastructure: implications for taphonomy and stable isotope analysis. *J. Archaeol. Sci.* 38:3358-72.

Taylor, C. R., Sober, A. J. (1996) Sun exposure and skin disease. *Annu. Rev. Med.* 47:181-91.

Uitto, J. (1986) Connective tissue biochemistry of the aging dermis. Age-related alterations in collagen and elastin. *Dermatol. Clin.* 4:433-46.

Urtiaga, M., de Pablo, N., Martinez, S. (2000) [Surveillance report on Diseases of Compulsory Notification in Navarra. 1999]. *An. Sist. Sanit. Navar.* 23: 293-9.

Varani, J., Perone, P., Griffiths, C. E., Inman, D. R., Fligiel, S. E., Voorhees, J. J. (1994) All-trans retinoic acid (RA) stimulates events in organ-cultured human skin that underlie repair. Adult skin from sun-protected and sun-exposed sites responds in an identical manner to RA while neonatal foreskin responds differently. *J. Clin. Invest.* 94:1747-56.

Varani, J., Schuger, L., Dame, M. K., Leonard, C., Fligiel, S. E., Kang, S., *et al.* (2004) Reduced fibroblast interaction with intact collagen as a mechanism for depressed collagen synthesis in photodamaged skin. *J. Invest. Dermatol.* 122:1471-9.

Varani, J., Spearman, D., Perone, P., Fligiel, S. E., Datta, S. C., Wang, Z. Q., *et al.* (2001) Inhibition of type I procollagen synthesis by damaged collagen in photoaged skin and by collagenase-degraded collagen in vitro. *Am. J. Pathol.* 158:931-42.

Varani, J., Warner, R. L., Gharaee-Kermani, M., Phan, S. H., Kang, S., Chung, J. H., *et al.* (2000) Vitamin A antagonizes decreased cell growth and elevated collagen-degrading matrix metalloproteinases and stimulates collagen accumulation in naturally aged human skin. *J. Invest. Dermatol.* 114: 480-6.

Varga, J. (2002) Scleroderma and Smads: dysfunctional Smad family dynamics culminating in fibrosis. *Arthritis Rheum.* 46:1703-13.

Varga, J., Rosenbloom, J., Jimenez, S. A. (1987) Transforming growth factor beta (TGF beta) causes a persistent increase in steady-state amounts of type I and type III collagen and fibronectin mRNAs in normal human dermal fibroblasts. *Biochem. J.* 247:597-604.

Verzijl, N., DeGroot, J., Thorpe, S. R., Bank, R. A., Shaw, J. N., Lyons, T. J., *et al.* (2000) Effect of collagen turnover on the accumulation of advanced glycation end products. *J. Biol. Chem.* 275:39027-31.

Von Gersdorff, G., Susztak, K., Rezvani, F., Bitzer, M., Liang, D., Bottinger, E. P. (2000) Smad3 and Smad4 mediate transcriptional activation of the human Smad7 promoter by transforming growth factor beta. *J. Biol. Chem.* 275:11320-6.

Wall, I. B., Moseley, R., Baird, D. M., Kipling, D., Giles, P., Laffafian, I., *et al.* (2008) Fibroblast dysfunction is a key factor in the non-healing of chronic venous leg ulcers. *J. Invest. Dermatol.* 128:2526-40.

Wang, F., Garza, L. A., Kang, S., Varani, J., Orringer, J. S., Fisher, G. J., *et al.* (2007) In vivo stimulation of de novo collagen production caused by cross-linked hyaluronic acid dermal filler injections in photodamaged human skin. *Arch. Dermatol.* 143:155-63.

Wang, Z., Boudjelal, M., Kang, S., Voorhees, J. J., Fisher, G. J. (1999) Ultraviolet irradiation of human skin causes functional vitamin A deficiency, preventable by all-trans retinoic acid pre-treatment. *Nat. Med.* 5:418-22.

Wlaschek, M., Tantcheva-Poor, I., Naderi, L., Ma, W., Schneider, L. A., Razi-Wolf, Z., *et al.* (2001) Solar UV irradiation and dermal photoaging. *J. Photochem. Photobiol. B* 63:41-51.

Yaar, M., Gilchrest, B. A. (1998) Aging versus photoaging: postulated mechanisms and effectors. *J. Investig. Dermatol. Symp. Proc.* 3:47-51.

Yaar, M., Gilchrest, B. A. (2007) Photoageing: mechanism, prevention and therapy. *Br. J. Dermatol.* 157:874-87.

In: Dermal Fibroblasts
Editor: Xiaowen Bai
ISBN: 978-1-62618-799-3

Chapter 4

Role of Dermal Fibrobasts in Post-Burn Hypertrophic Scarring

Jian Fei Wang,[1] Saad M. Alrobaiea[1] and Edward E. Tredget[1,2,*]

[1] Wound Healing Research Group, Division of Plastic and Reconstructive Surgery, Department of Surgery, University of Alberta, Edmonton, Alberta, Canada

[2] Critical Care Medicine, University of Alberta, Edmonton, Alberta, Canada

Abstract

Hypertrophic scar (HTS) following thermal injury is a dermal fibroproliferative disorder (FPD) that leads to considerable morbidity. The development of HTS involves numerous cell types and cytokines with dermal fibroblasts being a key cell. In this work, the properties of dermal fibroblasts in HTS were discussed. First of all, heterogeneous dermal fibroblasts were discussed, particularly focused on dermal fibroblasts from different depth. In normal skin, the properties fibroblasts isolated from different depth using dermatome are different. Cells from the deeper layers were observed to proliferate at a slow rate, but were morphologically larger. The cells from the deeper layers produced more TGF-β1 and CTGF. TGF-β1 producing cells in deeper layers were higher. Cells from the deeper layer tended to transform into myfibroblasts. Fibroblasts from the deeper layers were also found to produce more collagen, but less collagenase. Interestingly, cells from the deeper layers also produced more proteoglycn and versican, but less decorin. Therefore, fibroblasts from the deeper layers of the dermis resemble HTS fibroblasts, suggesting that the deeper layer fibroblasts may play a critical role in the formation of HTS. The immunoinflammatory property of dermal fibroblasts using TLR-4 as an example was also discussed. Furthermore, the phenotype of fibroblasts isolated

* Corresponding authors: Jianfei Wang and Edward E. Tredget, current address: Jianfei Wang, GlaxoSmithKline, R and D China, 4F, Building 1, 917 Halei Road, Zhangjiang Hi-Tech Park, Pudong, Shanghai 201203, China. Tel: (+86) 21-6159-0990 Fax: (+86) 21-6159-0704; E-mail: jianfei.j.wang@gsk.com; Edward E. Tredget, etredget@ualberta.ca.

from HTS was altered as indicated by changed molecular expression such as TGF-β and delayed apoptosis. Understanding heterogeneous property and the role of dermal fibroblasts in inflammatory and immune regulation in post-burn HTS and targeting stromal cell-mediated chronic inflammation offers the potential for novel exciting therapies - not only for dermal scarring, but possibly for other important types of FPD for which effective treatments are seriously lacking.

Keywords: Fibroblasts, hypertrophic scar, transforming growth factor-beta (TGF-β), inflammation, myofibroblast, regulation, apoptosis, Small leucine-rich proteoglycans (SLRPs)

Abbreviations

HTS	hypertrophic scar
FPD	fibroproliferative disorder
ECM	extracellular matrix
PMN	polymorphonuclear cells
TGF	transforming growth factor
TNF	tumor necrosis factor
MCP-1	monocyte chemotactic protein-1
ICAM	intercellular adhesion molecule
TLR	toll like receptor
BAMBI	bone morphogenic protein and activin membrane-bound inhibitor
RA	rheumatoid arthritis
ACAM	adipocyte adhesion molecule
SMA	smooth muscle actin
TUNEL	terminal deoxynucleotidyl transferase-mediated dUTP-FITC nick end labeling
FACS	fluorescent activated cell sorting
SDF-1	stromal cell derived factor
PGE2	Prostaglandin E2
UTS	Ultimate tensile strength

Introduction

Following thermal injury, many patients develop hyperprotrophic scarring (HTS) which is a common fibroproliferative disorder of the human dermis characterized by erythematous, raised, pruritic lesions of the healing skin, which is usually following thermal and other injuries that involve the deep dermis. [1] These lesions lead to scarring that compromises the appearance of healing skin and are commonly associated with contractures that limit movement and function of involved joints and facial features. As such, HTS is the principal factor that contributes to the prolonged and often uncomfortable rehabilitation period for thermally injured patients, particularly those who have survived large life-threatening injuries, children, and individuals of dark-skinned races, in whom such scarring occurs more commonly. [1] The undesirable physical properties of HTS tissue can be attributed to the presence of a large amount of extracellular matrix that is of altered composition and organization, compared to normal dermis or mature scar.

This matrix is the product of a dense population of fibroblasts, maintained in a hyperactive state by inflammatory cytokines such as TGF-β and other factors, some of which may be physical in origin. [2] Although the molecular and cellular events that lead to HTS have been extensively studied, the pathogenesis of this condition is still not well understood, making treatment difficult.

Key cells involved in HTS are the dermal fibroblasts. Dermal fibroblasts are a dynamic and diverse population of cells whose function in skin and HTS in many respects remain unknown. [3] Normal adult human skin contains at least three distinct subpopulations of fibroblasts, which occupy unique niches in the dermis. Fibroblasts from each of these niches exhibit distinctive differences when cultured separately. [3]

Specific differences in fibroblast physiology are evident in papillary dermal fibroblasts, which reside in the superficial dermis, and reticular fibroblasts, which reside in the deep dermis.

Another subpopulation of fibroblasts is associated with hair follicles [4] and engages in fibroblast-epidermal interactions during hair development and in interfollicular regions of skin.

Dermal Fibroblasts Are Heterogeneous and Deep Dermal Fibroblasts Contribute to More Scarring

Studies have shown that dermal fibroblasts are an extremely heterogeneous multifunctional cell population and play an important regulatory role in wound healing. Fibroblasts cultured from the papillary and reticular dermis have different characteristics including difference in their proliferation kinetics, rates at which they contract type I collagen lattices, production of the proteoglycans decorin and versican, and their size and packing density in confluent cultures. [3-7]

Fibroblasts associated with hair follicles also show a distinctive characteristic such as different amount of protein production and are different from the cells in the papillary and reticular dermis. [8, 9]

Studies have demonstrated that dermal fibroblasts present even within the same tissue sample are heterogeneous and their microenvironmental localization might be one of the main reasons for fibroblast heterogeneity. [10]

Studies have also demonstrated that the superficial papillary dermis contains a heterogeneous population of fibroblasts, and proposed that diverse subpopulations of fibroblasts are required to interact in both homeostatic and pathological situations in the skin. [11]

The association between scarring and depth of dermal injury is clinically recognized and is evident from lasers used to treat superficial vascular malformations and facial resurfacing. [12, 13] that depth of the injury, as well as the site of the wound, dictate the outcome of wound appearance, where superficial burn will heal quickly without scar. In a porcine skin wound model, a deep, partial thickness dermal burn resulted in HTS. [14] Dunkin et al. quantified the association between scarring and the depth of dermal injury in human volunteers using a novel jig saw scratch/incision approach, where they found that there was a threshold depth of dermal injury at which scarring develops. [15]

These studies imply that deeper dermal fibroblasts may play a critical rolc in HTS.

More recently study shows that dermal fibroblasts are highly heterogeneous and deep dermal fibroblasts may play a critical role in HTS. Different layers of skin fibroblasts were isolated and cultured by using dermatome separation. [16] Dermal fibroblasts from deeper layers were observed morphological larger and proliferated slower than those of the superficial layers. In addition, deep dermal fibroblast produced less collagenase but more collagen, and they produced less decorin but more versican.

Moreover, cells from the deeper layers produced more TGF-β1 and α-SMA compared to cells from more superficial layers. This study indicates that the phenotype of dermal fibroblasts from the deeper layers resemble those from HTS and strongly suggests that deep dermal fibroblasts may play a critical role in HTS. [16-18]

Fibroblasts from Deep Dermis Produce More TGF-B_1 [16]

The development of HTS is often linked to the over-expression of TGFβ1. [19] Elevated levels of TGF-β1 are found in the serum of recovering burn patients, [19, 20] and HTS tissues stain more intensely for this cytokine than do normal skin samples or mature scars. [21] TGF-β1 is therefore a key cytokine in wound healing and hypertrophic scarring. [1]

TGF-β1 is produced by all the major cell types participating in wound repair, including T-lymphocytes, macrophages, smooth muscle cells, endothelial cells, fibroblasts, epithelial cells, and fibrocytes. [22, 23] Fibroblasts are the predominant cell type involved in healing wounds and may provide much of the TGF-β1 in an active HTS, especially as its synthesis can be auto-induced in these cells. [24]

We previously reported that fibroblasts from HTS produced significant amount of TGF-β1 protein compared to fibroblasts from paired normal fibroblasts. [25] We recently reported that fibroblasts from deeper dermal layers progressively produced increasing amounts of TGF-β1 compared to superficial layers. Moreover, a large proportion of TGF-β1 producing fibroblasts was observed in the deeper layers of the dermis. This data suggests that fibroblasts from the deeper dermal layers resemble fibroblasts in HTS in terms of TGF-β1 production and as such may contribute substantially to the cell population in HTS. [16]

Fibroblasts from Deep Dermis Produce More Connective Tissue Growth Factor (CTGF) [16]

Previous studies have shown that TGF-β1 and CTGF coordinately function together to regulate wound healing. [26] CTGF exhibits numerous biological properties that are of potential importance in the wound healing response, including stimulation of cell proliferation, cell adhesion, chemotaxis, angiogenesis and production of ECM components. [27, 28] In addition, like TGF-β1, CTGF has been linked to several fibrotic disorders. [29] CTGF mRNA expression was reported to increase 20 fold in isolated fibroblasts from HTS compared to normal fibroblasts. When stimulated with TGF-β1, fibroblasts from HTS were seen to increase CTGF mRNA expression more than 150 fold compared with normal fibroblasts. [30] In the study that fibroblasts isolated from different dermal layers, it was indicated that fibroblasts derived from the deeper dermis demonstrated a significantly increased CTGF mRNA levels compared to cells from superficial layers. [16] This suggests that the fibroblasts from the deeper dermal layers resemble HTS cells in terms of CTGF production. In addition, mRNA levels for CTGF in TGF-β1 neutralizing antibody treated deep dermal cells is significantly downretulated compared to IgG isotype treated cells, suggesting difference in CTGF expression in fibroblasts from different layers may at least partially result from TGF-β1 and be secondary results.

Fibroblasts from Deep Dermis Transform More Myofibroblasts [16]

An important part of wound healing is wound contraction and myofibroblasts are believed to play a crucial role in this process. [31] Myofibroblasts are characterized by the expression of α-SMA and are thought to be derived from the fibroblast population under profibrotic growth factor stimulation. [32] Following skin injury, fibroblasts differentiate into contractile and secretory myofibroblasts that contribute to tissue repair during wound healing. [31] However, in HTS, due to an increased number of myofibroblasts, contraction and ECM protein secretion become excessive and that severely impairing skin function. [31] Myofibroblasts are characterized by the expression of α-SMA and are thought to be derived from the fibroblast population under stimulation by profibrotic growth factors such as TGF-β1. [33] In the following study, using paired fibroblasts, we have shown an increase in α-SMA expression in fibroblasts derived from HTS compared to cells from normal skin.

Interestingly, fibroblasts derived from the deeper dermal layers produced more α-SMA protein compared to cells from the superficial layers, consequently, the percentage of α-SMA producing fibroblasts was higher in the deeper layers. When cells were incorporated in collagen lattice gel, fibroblasts from the deeper layers contracted the lattice more than cells from the superficial layers. Taken together, this data strongly suggests that deep layer dermal fibroblasts resemble the phenotype of HTS fibroblasts in terms of α-SMA expression.

Fibroblasts from Deep Dermis Produce More Collagen and Less Collagenase [16]

Under physiological conditions, there is a balance between the synthesis and degradation of collagen in human dermal tissues [34] and we previously reported that this balance was lost in HTS formation. [35]

Extensive thermal injuries often disrupt this balance and lead to abnormalities in collagen metabolism, a characteristic feature of several forms of fibrotic conditions. [36, 37] HTS is characterized by increased collagen deposition. [38] Thus, alteration in ECM composition is believed to be partially responsible for the undesirable physical properties of HTS tissue.

We previously reported evidence that HTS fibroblasts expressed much less collagenase mRNA than normal cells obtained from the same patients, but instead expressed more collagen type I and III mRNA. [35] This differential synthesis of collagen and collagenase in HTS appears to be of an important feature which leads to excessive accumulation of collagen in this fibrotic condition. In the following study, dermal fibroblasts from deeper layers were observed to produce more collagen but less collagenase, resembling the fibroblast phenotype from HTS. In addition, our results revealed that increased collagen production observed in deeper layer skin fibroblasts was accompanied by the up regulation of Hsp47 expression, a collagen chaperone. As Hsp47 plays an important role in collagen biosynthesis, [39] this feature of fibroblasts from the deeper dermal layers is consistent with increased synthesis and altered assembly of collagen. Thus, in terms of ECM production, fibroblasts derived from deeper dermal layers produced more collagen and less collagenase than cells from superficial layers, which further suggests that fibroblasts from deep dermal layers resemble HTS fibroblast phenotype.

Fibroblasts from Deep Dermis Make Less Decorin but More Versican [16]

Proteoglycans influence physical properties of connective tissues and influence cellular activity. The morphology of collagen fibrils and their organization are profoundly affected by the nature and amounts of proteoglycans present in the connective tissue. [2]

In HTS after thermal injury, the content of proteoglycan has been reported to be altered as compared to normal skin. Decorin, a small dermatan sulfate proteoglycan decorin, is the major proteoglycan found in normal dermis. HTS contains on average only 25% of the amount of decorin found in normal dermis.

It also contains 6-fold higher concentrations of the large proteoglycan, versican. [40] Moreover, fibroblasts cultured from post-burn HTS contain less decorin mRNA and synthesize less of the protein than do normal fibroblasts. [41] In the following study, fibroblasts from the deeper dermal layers produce less decorin compared to cells from superficial regions and there is a decreasing percentage of decorin producing cells with increasing dermal depth. In contrast, higher percentages of versican producing cells were observed with increasing dermal depths. This again suggests that fibroblasts from the deeper dermal layers are more similar to HTS fibroblasts in terms of proteoglycan production than are superficial fibroblasts.

When fibroblasts receive stimulation in the injured tissue environment, cell spreading, nuclear extension, DNA synthesis, and protein production will result in an increase of cell size. Thus stimulated fibroblasts demonstrate increased size and slower rate of proliferation.

Moreover, this increased size of fibroblasts is likely associated with the increased capacity of cells to synthesize ECM. In our study, we have compared the cell size of normal and HTS fibroblasts from the same patients and found that HTS fibroblasts were much larger than those from normal skin.

Furthermore, cells from the deeper dermal layers demonstrate significant increase in and decreased proliferation rate relative to superficial layers. This further suggests that the phenotype of cells from deeper dermal layer in terms of size and proliferation is similar to that of HTS fibroblasts.

Deep Dermal Fibroblasts Refractory to Migration and Decorin-Induced Apoptosis [17]

Fibroblasts are abundant cells of connective tissue and they have an extremely important role in wound healing. They continuously produce components of the extracellular matrix that maintain the physical integrity of the connective tissue.

There are two characteristics features of fibroblast activation, which are the phenotypic modification and substantial proliferation. [42] Many cells and cytokines are contributing in hypertrophic scar formation; however the dermal fibroblasts are key cells in this process. Hypertrophic scar is a dermal form of fibroproliferative disorders and persisting of myofibroblasts in the tissue is responsible of the pathological fibrosis by increasing matrix synthesis and contraction of the tissue furthermore there are more fibroblasts and myofbroblasts in hypertrophic scar tissues in comparison with the normal skin and mature scars which make an evident of their role in fibrosis. [16, 43] Within the normal skin there are significant differences between the superficial dermal fibroblasts (SF) and the deep dermal fibroblast (DF), which also known as reticular fibroblasts in terms of molecular and physical characteristics. Deep fibroblasts proliferate in slower rate and produce more TGF-β1 and also found to produce more collagen but less collagenase in comparison to the superficial fibroblast.

In addition, deep dermal fibroblast within normal skin produce more α-SMA and more α-SMA-positive cells which contract collagen to more extent, indicating the fibrotic feature of deep dermal fibroblasts.

Moreover, injuries to the superficial dermis heal with little or no scar and the deep wounds usually lead to development of hypertrophic scar. [18] Myofibroblasts are crucial cells for dermal wound healing by inducing contraction and granulation tissue formation, which are two important events in the process of wound repair. Myofibroblasts are observed transiently in the wound during normal healing process and their presence indicates active contraction phase of the healing wound.

After the contraction phase has stopped there will be no myofibroblast detected in normal scarring this is mediated by apoptosis of the cells between the period of the granulation tissue to scar formation. [32, 33] Decorin and small leucine-rich proteoglycan (SLRP) may be involved in cell apoptosis and reducing granulation tissue hyper cellularity in addition to their role in collagen fibrillogenesis and TGF-β1 [44] which was observed in oral mucosal wound of decorin knockout mice when fibroblast increased by twofold in the oral wound. [45] Repithelialization and granulation tissue formation require proliferation and migration of cells from the wound edge to fill the wound site and during the inflammatory phase of wound repair fibroblast migrate to the wound region to replace the blood clot with collagen and fibronectin which is synthesized by fibroblast mediates the adhesion of the cells to the connective tissue matrix. [46] Therefore, the wound healing outcome may be altered by the differences in migratory behavior and adhesion feature of fibroblast in different skin layer.

In the following study, the investigator studied the cell viability, migration, adhesion and decorin-induce gene expression of apoptotic factors in superficial and deep dermal fibroblast obtained from normal skin. Following informed consent from patients who underwent elective abdominoplasty surgery, superficial and deep dermal fibroblasts were isolated and used in this study using an electric dermatome set at approximately 0.5 mm and the skin tissue samples were horizontally sections into five layers (referred to layers 1 to 5) as the superficial layer is [layer 1] and the deep is [layer 5] and to study the differences of the physical features of the superficial fibroblast and deep fibroblast, cells were isolated from the dermis as layer 1 (1-100 μm) and layer 5 (400-500 μm).

Cell migration assay was performed on SF and DF to test the difference in fibroblast migration from each layer of the skin, which shows that the number of migrating cells from SF is higher than the DF. In addition, the SF appears to be smaller and bipolar in shape.

They stretched in one direction where cell processes extend from one cell to another indicating migration compared with the DF which appears to be large, flat and having multiple cell processes extended from many cell poles, indicating a stationary behavior. Cell adhesion assay was performed on fibroblast taken from the skin of different subjects that interestingly showed no difference of adhesion of SF and DF to fibronectin.

Apoptosis is critical step in wound repair that lead to decrease the number of fibroblast in granulation tissue which in turn reduce the scar formation by lowering the deposition of ECM. Lack of apoptotic stimuli to some extent my play a role in HTS formation. Decorin is a proteoglycan that has been shown to induce cell growth suppression and apoptosis in many cell types by interacting with cell surface receptors. [47, 48] Hypertropgic scar has been shown to have reduced amount of decorin. [41] Decreased amount of decorin may be contributing in hypertrophic scar formation in many ways. Lack of decorin resulted in thick unorganized collagen fibers. [49]

In addition, the activity of fibrotic TGF-β1 increased if the decorin level was reduced. Furthermore, the lack of decorin in hypertrophic scars led to the decrease of the reduction of myofibroblasts in granulation tissue that normaly happened during wound healing. [50, 51] Taken together these findings suggest the role of reducing the amount of decorin in HTS formation.

Examination by flow cytometry shows that compared with control cells, treatment of the fibroblast with 50 mg/ml decorin for 24 hours resulted in the increase of fibroblast cell apoptosis in SF and DF about 5.1 folds and 4.8 folds, respectively.

Comparing with DF, SF had higher number of apoptotic cells with decorin treatment and higher trypan blue staining, which represent the number of non-viable cells after treating the cells with decorin. These results suggest that DF may have some resistance to the apoptotic effect of decorin in comparison with SF. Quantitative RT-PCR also performed to analyze the mRNA expression of the apoptotic genes that can be induced by decorin and showed that mRNA level of histone-1, caspase-1, caspase-8 and p53 increased after treating the cell with 100 μg/ml of decorin for 60 minutes in SF compared with DF; however, decorin did not induce any significant changes in the expression Bax or Caspase-3. Thus, the role of decorin in reducing cells at specific time during wound healing needs further investigations.

Reduced Decorin, Fibromodulin, and TGF-B3 in Deep Dermis [18]

Decorin and fibromodulin are members of the small leucine-rich proteoglycan (SLRP) family that present in the extracellual matrix (ECM) in which they interact with collagen modifying their arrangement and deposition in the ECM. SLRP also interact with cytokines and growth factors like TGF-β and interfere with their activity. [52, 53] The level of TGF-β1 has been shown to be increased in HTS tissue [19] and it has been reported that their activity can be inhibited after binding to SLRP specially decorin and fibrmodulin in vitro and in vivo. [53, 54] these findings suggest that decreasing the level of decorin and fibromodulin during wound healing my result in HTS formation.

There are three isoforms of TGF-β, which are TGF-β1, TGF-β2 and TGF-β3 [55] which are encoded by different gens and all of them are present in wound healing process and have overlapping their biological functions. [56, 57]

TGF-β1 expression is decreased by the influence of the nonfibrotic cytokine TGF-β3 and the balance between them may have an important role in scar formation. [58, 59] High expression of TGF-β3 was seen in fetal tissue fibroblast which heals without scaring. [60, 61] In addition, the ratio of TGF-β3 to TGF-β1 was higher in the non-scarring gingival wound model. [62] These observations suggest that decreasing the ratio of TGF-β3 to TGF-β1 may play a role in the development of HTS.

In the following study, the investigators aim to identify the expression and localization of factors that regulate wound healing including decorin, fibromodulin, TGF-β1 and TGF-β3 in an experimental human scratch wound and also from fibroblast from superficial and deep layer of normal skin. A specially designed jig using a number 11 scalpel blade was created after the original description of human dermal scratch model by Dunkin et al. [15] and after obtaining an informed written consent from a burn patient an incisional wound was made using the jig located midway between anterior superior iliac spine and greater trochanter under general anesthesia in sterile condition in the operating room at the time of split thickness skin graft harvesting. The incision was 6 cm long, 0 mm to 0.75 mm in depth at one end (superficial wound) and 0.76 mm to 3 mm at the other end (deep wound). The wound healing progression was observed. Around 70 days post wounding the superficial wound healed with minimal scaring (SWS) while the deep wound at the other end healed with red, raised and hard scar (DWS) that also confined to the boundaries of the wound, which grossly resembled HTS. In addition the DWS, we also observed the thicker epithelium, hypercellularity, hypervascularity and bulky collagen bundle fibers compared with SWS at microscopic level, indicating that the DWS develop microscopic features of HTS.

Immunofluorescence staining was used to analyze the level of decorin and fibromodulin in the wound. The level of staining intensity of decorin and fibromodulin was lower in DWS compared to the one in SWS. However, there were no any differences in the other structure- and function-related SLRPs, biglycan and lumican. Immunoreacrivity of normal skin fibroblasts from different layer showed the decreased activity of decorin and fibromodulin in deep layer (L5) fibroblast compared with the superficial (L1) fibroblasts in addition to the downregulation of decorin and fibromodulin expression in L5 fibroblasts which was analyzed using western blotting.

Staining intensity for TGF-β1 was higher in DWS compared to SWS but it was invers regarding the nonfibrotic TGF-β3 which was decreased in DWS in comparison to SWS. In addition the area that lays directly below the basement membrane of the epithelial cells in SWS shows numerous TGF-β3 which was not seen in DWS. Furthermore immunoreactivity of TGF-β1 in the cultured fibroblast from normal skin tissue from different layer was higher in L5 compared to L1 in contrast to TGF-β3 immunoreactivity, which was higher in L1 fibroblasts.

TGF-β1 needs to bind to its type I and type II receptors to start unique down stream signaling pathway. HTS tend to develop if there is increaseed expression of TGF-β receptors and extended presence of TGF-β1. RT-PCR, western blotting and immunofluorescence staining were used to measure the expression of TGF-β receptor II that has high binding affinity to TGF-β1. The number of type II TGF-β receptor and their immunoreactivity in L5 fibroblasts was higher than L1 fibroblasts evidenced by twofold higher level of TGF-β type II protein expression in L5 fibroblasts than L1 fibroblasts.

Furthermore, in this study the expression of the Thy-1 antigen was investigated. Thy-1 is cell surface antigen that has been reported to play a role in differentiation of fibroblast into myoibroblast by TGF-β1 and the responsiveness of the fibroblast to that stimulation depends on whether they are Th-1 negative or Th-1 positive fibroblasts. Th-1 negative fibroblasts showed a greater response to TGF-β fibrogenic stimulation. [63-65] Using flow cytometery, the investigators were able to show that the DF lack or weakly express Th-1 in contrast to SF where the Th-1 is strongly expressed. Therefore this suggests that the formation of HTS after deep dermal wounds may be associated with the lack of expression of Th-1 surface antigen in DF.

Taken together, the data from this investigation suggest that HTS formation may be reduced by the increased expression of TGF-β3 and SLRPs, decorin, fibromodulin, Th-1 surface antigen during wound healing, wheter it is true remains more functional studies.

Differential Collagen-Glycosaminoglycan Matrix Remodeling by Superficial and Deep Dermal Fibroblasts [66]

Skin as the largest organ of the humans, is more prone to external injuries and trauma which varies from simple cut wounds to more extensive deep dermal wounds as third degree burns thus, the treatment of these wounds usually relay on their depth, severity and the anticipated outcomes of it as some these wound may be treated with simple wound care and closure which used in superficial skin cut injuries where others may requires skin autografts like in deep skin lesions or burns.

In some sever cases of extensive skin loss or third degree burns suitable donor site for autografting is not available therefore, cultured skin substitutes (CSS) is the most favorable option of treatment in these cases. CSS containing autologous keratinocytes and fibroblasts seeded on an artificial extracellular matrices (ECM) made of collagen-glycosaminoglycan (C-GAG) shows a great potential to be used in skin repair since more promising clinical result for wound care have been reported by using it. [67]

A major undesirable outcome of post-burn wound healing is HTS formation, which is due to excessive accumulation of ECM that has different organization and composition compared with normal skin. Two-thirds of patients who have burn injuries develop HTS that result in long unpleasant rehabilitation period, considerable disfiguring, and functional disabilities. [67] There relation of the depth of the wound an the formation of HTS is clinically observed and shows that the superficial wounds usually heals with no or minimal scarring in contrast to deep wounds which are more likely to develop elevated, red and hard scars that are confined to the original injury borders resembling HTS. [15, 18, 68]

Fibroblasts are key cells in wound healing and there are three subpopulation of fibroblast that shows various features when cultured separately. [69] Two of these fibroblast sub-populations are located in the papillary and reticular dermis which known as superficial (SF) and deep fibroblasts (DF), respectively; however the third sub-population is associated with hair follicles. It is reported that there are differences in physical and biochemical features including size, packing, proliferation rate, production of collagen and type I and III procollagen between superficial and deep fibroblasts. [3, 6, 7, 10] In addition, DF have been reported to produce more TGF-β1, Type I collagen and the large proteoglycan versican but less decorin compared to SF. Also, it was shown that DF may be similar to fibroblasts found in HTS. [16] Thus, in the following study the investigators aimed to analyze the difference in the remodeling of SF and DF on C-GAG matrices since the current CSS are prepared of different population of dermal fibroblast C-GAG matrices.

C-GAG matrices containing SF and DF separately were analyze to have an idea about the molecular basis and functional differences of fibroblasts and their participating in HTS formation. Fibroblasts used in this study were taken from abdominal tissues from three patents that underwent elective abdominoplasty surgery after taking an informed written consent from them.

Then the skin tissue was separated to five different layers (L1 to L5) where the superficial layer is L1 and the deep layer is L5 using a dermatome that seated approximately at 0.5 mm. To set the extent of the contractions mediated by fibroblasts, C-GAG matrices cultured independently with SF or DF were analyzed at different time points. The original area before seeding of fibroblast was considered as 100% and the percentage of the contraction was calculated at each time point with respect to this original area and C-GAG without seeding of fibroblast was used as controls. This shows that the matrices without fibroblasts did not contract at all time points where the matrices with fibroblasts did. Interestingly, the matrices seeded with DF contracted more than the matrices with the SF and this difference was observed on days 7, 14 and 21 of culture. Furthermore, the stiffenss and Ultimate tensile strength (UTS) using tensile tester of the C-GAG matrices were studied and showed that the control matrices lacking fibroblasts had significant lower stiffness and UTS compared with matrices with fibroblasts. It was also observed that the C-GAG matrices seeded with DF were stiffer than the one seeded with SF while the matrices seeded with SF had higher UTS than the matrices with DF. This stiffness in DF may be due to their production of higher amount of collagen. The differences in proteoglycan expression by SF and DF may contribute in the observed difference of UTS.

Quantifying hydroxyproline level that present in the matrices and in conditioning medium by mass spectrometry allows determining the amount of collagen production by fibroblasts. The results showed that the conditioning medium taken from matrices with SF had a significant lower amount of hydroxyprolin than that obtained from DF at days 14 and

21 of culture. In addition, the matrices with DF had higher hydroxyprolin level than matrices containing SF at day 4 and 14 of culture.

Histological analysis of the matrices reveled that the matrices with DF stained darker with hematoxylin than the matrices with SF, refering to presence of higher amount of ECM and the control matrices without fibroblasts showed lower hematoxylin staining than the matrices with fibroblasts.

To investigate the extent of differentiation of fibroblasts to myofibrblasts that mediates wound contraction, the matrices were subjected to α-SMA staining at different time points of culture. The staining found to be maximum at day 14 for both SF and DF matrices; however, the matrices containing DF showed more α-SMA staining than the matrices congaing SF, suggesting that more DF differentiate into myofibroblasts in comparison to SF.

Expression of different gene was analyzed in this study using quatntative RT-PCR to understand the molecular basis of the role of SF and DF in HTS. The data showed that DF had higher expression of CTGF, TGF-β1, HSP-47, osteopontin, angiotensin-II, peroxisome prlifrator-activated receptor (PPAR)-β than SF; however DF showed lower expression of tumor necrosis factor-α, PPAR-β/γ, PPAR-γ decorin and fibromodulin than SF.

In conclusion, SF and DF differently remodeled the C-GAG matrices. DF led to a greater stiffness, contraction of the matrices, and a decrease in tensile strength than SF, suggesting that using CSS with specific fibroblast sub-population (e.g., SF) rather than the one with mixed population of fibroblasts that have been used currently may be more helpful and have favorable outcome in wound healing in extensive deep wounds.

Dermal Fibroblasts in HTS Are of Immunoinflammatory Property

Skin wound healing is intimately associated with the onset of acute inflammation and the arrival of neutrophils. [70] Circulating monocytes arrive shortly after neutrophils at the wound site, where they are activated and subsequently transform into macrophages. [70, 71] T lymphocytes normally infiltrate wounds later and also play a significant role in wound healing. [72]

Overall depletion of all subsets of T cells results in impaired wound healing. [72-74] However, each subset appears to have different effects on wound healing. [75] Thus, the interaction of various infiltrating inflammatory cell types with resident skin cells can profoundly affect the healing process, dictating the outcome.

Fibroblasts were, for a long time, considered to be responsible only for the production of extracellular matrix components. This assumption has been changed with the emergence of evidence showing that fibroblasts can also participate in immunological responses in direct response to proinflammatory signals in areas such as: the regulation of normal barrier function of the epithelium remodeling of infected tissue [76, 118, 119] and regulation of the behavior of infiltrating leukocytes to sites of inflammation. [77, 78]

Evidence shows that fibroblasts regulate immune/inflammatory responses through TLR4 activated by LPS through adaptor molecules like MyD88, leading to NF-κB and MAPK activation, cytokine gene transcription and co-stimulatory molecule expression such as CD40 resulting in inflammation. This study explored the possible role of TLR4 in HTS. Paired normal and HTS tissues from burn patients were collected and dermal fibroblasts were then isolated and cultured.

Immunohistochemical analysis of these tissues demonstrated increased staining intensity of TLR4 in HTS tissue. Assessing three pairs of fibroblasts by quantitative RT-PCR demonstrated that mRNA levels for TLR4 and MyD88 in fibroblasts from HTS were increased significantly compared with fibroblasts from normal skin. Flow cytometry analysis indicated increased TLR4 expression in fibroblasts from HTS compared with normal skin. ELISA analysis demonstrated protein levels for cytokines PGE2, IL-6, IL-8 and MCP-1 were significantly increased in fibroblasts from HTS compared to normal skin. When normal fibroblasts were stimulated with LPS, significant increase in mRNA and protein levels for MyD88, IL-6, IL-8 and MCP-1 was detected. However, when cells were transfected with MyD88 siRNA, then stimulated with LPS, a significant decrease in mRNA and protein levels for these molecules compared to LPS stimulated fibroblasts was detected. In comparison, a scramble siRNA transfection did not affect mRNA or protein levels for these molecules. Results demonstrate LPS stimulates proinflammatory cytokine expression in dermal fibroblasts and MyD88 siRNA eliminates expression. Therefore, controlling inflammation and manipulating TLR signaling in injured skin cells may result in novel treatment strategies of HTS and other fibroprolifrative disorders.

Delayed Fibroblast and Myofibroblast Apoptosis HTS

One of the features of hypertrophic scar is contracture and myofibroblasts are transiently found at the site of tissue injury. [33] By secreting extracellular matrix proteins and by promoting the contraction of the granulation tissue through the expression of α-SMA, these cells are believed by some investigators to be essential for wound closure.

Increased numbers of myofibroblasts were found in hypertrophic scar. [79, 80] Delayed apoptosis of myofibroblasts has long been suggested in hypertrophic scar. The protein of bcl-2 proto-oncogene that was reported to be able to protect cells from apoptosis, was strongly up-regulated in hypertrophic scar, [81] while the death receptor ligand Fas,was down-regulated both in hypertrophic scar and in the HTS isolated fibroblasts. [82] The transcription factor p53, regulating the expression of genes involved in cell cycle arrest or apoptosis in response to genotoxicity or other cell stress, was significantly down-regulated in hypertrophic scar. p53 can bind to MMP-2 (matrix metalloproteinase 2) promoter and up-regulate its activity, thus decreasing net deposition of collagen fibroblasts.

More recently direct evidence regarding delayed apoptosis was reported by Moulin et al. who examined anti-apoptotic and pro-apoptotic gene expression. [81] Isolated fibroblasts from normal healing wounds showed higher rates of apoptosis than those from hypertrophic scar; meanwhile, low levels of the anti-apoptotic proteins Bcl-2 and BclxL were detected in normal wound myofibroblasts. These cells also showed an increase in the level of the pro-apoptotic Bax when compared to normal skin fibroblasts and myofibroblasts from hypertrophic scar. Myofibroblasts from hypertrophic scar showed a higher level of Bcl-2 compared to fibroblasts but no difference in the Bax or BclxL levels. After serum starvation, wound myofibroblasts showed an increased apoptotic rate, whereas hypertrophic scar myofibroblasts and fibroblasts did not show any difference. Anti-Fas treatment did not modify the levels of apoptosis but strongly increased the growth of hypertrophic scar myofibroblasts as compared to normal wound myofibroblasts. More recently, Linge et al. observed that hypertrophic scar fibroblasts exhibit resistance to a specific form of apoptosis elicited by contraction of collagen gels, and that this phenomenon is dependent on excess activity of cell surface tissue transglutaminase. [84]

Taken together, these data have confirmed the hypothesis of defects in apoptosis and growth during pathological scar formation, resulting in delayed disappearance of fibroblasts and myofibroblasts.

Fibroblasts from HTS Tissues Produce More TGF-Β1 Than Normal Skin Cells

TGF-β1 is the prototypic member of a structurally related protein family that is involved in many proliferative, inductive, and regulatory processes. [85] TGF-β is one of the most prominent profibrotic growth factors in wound healing. [86] TGF-β1 is overexpressed in hypertrophic scar. [1, 87] TGF-β has been linked to HTS and keloid formation in a number of ways. Strong and persistent expression of TGF-β and its receptors has been shown in fibroblasts of postburn HTS. As well, overexpression of TGF-β1 and β2 has been found in keloid and keloid-derived fibroblasts, with significantly lower TGF-β3 mRNA expression. [88] It has also been identified that the serum of recovering burn patients contains levels of TGF-β1 about twice those in control patients. [89] Furthermore, the ontogenetic transition from scarless fetal wound healing to adult scarring is believed to be TGF-β-dependent. TGF-β expression is only transiently expressed in scarless fetal wounds at a lower level that in adult wound healing. TGF-β receptors 1 and 2 are also expressed less. [90] Perhaps the most compelling evidence for a causative role of TGF-β in producing scar, lies in the finding of scar formation in fetal wounds treated with TGF-β. Evaluation of TGF-β1 gene expression in HTS and normal skin tissue indicates that mRNA level for TGF-β1 in HTS tissues is significantly higher than in normal skin tissues. Further analysis TGF-β1 level in fibroblasts derived from tissues indicates that both gene expression and protein levels for TGF-β1 in cells derived from HTS are much higher than in cells from normal skin tissues. Taken together, targeting TGF-β1 or controlling their level of expression in wounds healing may be one of the treatment approaches to minimize abnormal scar and HTS formation, which warrant further investigations.

Dermal Fibroblasts from HTS Produce Less Small Leucine-Rich Proteoglycans, Decorin and Fibromodulin [91]

Wound healing is a complex physiological process that requires interaction of many cells such as platelets, fibroblasts, epithelial and endothelial cells with different ECM molecules, cytokines and growth factors to retain the integrity and normal shape and function of the skin.

Failure of this process due to an extensive skin loss will lead to chronic non-healing wound and abnormal scaring. HTS is a form of abnormal scars that characterize by disorganize collagen deposition. [2, 70] Small leucine-rich proteoglycans (SLRPs) are ECM molecules that play an important role in wound healing by regulating collagen fibrillogenesis and inhibiting the activity of TGF-β. [52, 62, 92]

Decorin, fibromodulin and biglycan are members of the SLRPs family that can that have been reported to bind and decrease activity of TGF-β1 activity in vitro and in vivo. [53, 54, 93] Fibromodulin overexpression mediated by adenoivirs resulted in improving wound healing in incisional wounds and minimizing scarring. [94] In addition cultured fibroblasts taken from HTS have been reported to decrease collagen and TGF-β synthesis after treating it with decorin. [51, 95] The following study aimed to show and analyze the difference in the expression, cell-association and localization of SLRPs between normal skin and HTS. Following an informed consent taking from patients who were treated in the University of Alberta Hospital, Firefighters' Burn Treatment Unit and recovering from deep thermal injury that showed clinical features of HTS. Tissue samples from HTS where taken with site matched normal skin tissue samples from the same patient as controls. Histological analysis of these specimens showed that the HTS tissue samples had thicker epithelium and higher cell density than normal skin samples. The collagen fibers were thin and organized in basket-weave orientation in normal skin while the collagen in HTS samples had thick fibers and parallel noodle-shaped orientation in one direction indicating contraction. In addition the HTS reveled remarkable hypervascularity and hypercellularity compared with normal skin tissue. Using immunofluorescence staining, the SLRPs immunoreactivity showed that decorin reactivity was lower in HTS than the normal skin. Interestingly, in normal skin, decorin immunoreactivity was lower in deep dermal layer compared with the superficial dermal layer. The thick and dense collagen deposition was observed in deep dermis, suggesting a relation between a higher decrease of decorin expression and higher deposition of the thick collagen fibers. Fibromodulin and lumican staining were higher in normal skin than HTS in the contrary to biglycan, which stain more in HTS than in normal skin. Western blotting analysis showed that the expression of decorin and fibromodulin were decresed in HTS compared with normal skin. No significant difference was observed betewwn normal skin and HTS in terms of biglycan and lumican expression. Quantitative RT-PCR demonstrated that the expression of decorin and fibromodulin mRNA was lower in HTS compared with normal skin whereas biglycan and lumican showed higher mRNA expression in HTS. The data seen above give a proposition about the SLRPs and their activity in wound healing and how the decreased decorin and fibromodulin participate in HTS formation, the molecular and cellular role of SLRPs in wound healing need further functional studies and may lead to a novel modality in preventing and treating HTS.

Dermal Fibroblasts from HTS Produce Less Nitric Oxide Synthase Expression and Nitric Oxide [96]

Nitric oxide (NO) is crucial cellular signaling molecule in human body that involves in many physiological and pathological processes such as controlling neovascularization and vascular tone, [97] airway tone, [98] gastrointestinal peristalsis, [99] and insulin secretion. [100] It has an antimicrobial activity. [101]

Furthermore, it is involved in the development of the nervous system and acts as neurotransmitter. [102] NO also contributes in immune system regulation [103] and it has an anti-proliferative functions. The production of NO is controlled by nitric oxide synthase (NOS) which has three isoforms, two of them are mainly expressed and their activity depend on elevation of intracellular Ca^{2+} and extrinsic calmodulin which are endothelial eNOS and neuronal nNOS. The third isoform is not expressed in normal conditions and it is activity is independent from the level of intracellular Ca^{2+} and it shows to be induced in many cells by inflammatory cytokines and bacterial lipopolysaccharide (LPS) which is iNOS [104, 105, 106].

HTS is a dermal form of fibroprolifrative disorders that characterize clinicaly by hyperemic, raised scar with altered skin sensation that is confined at the original injury borders [23] and at cellular level by excess deposition of thick, dense collagen fibers produced by fibroblasts. [107]

NO with its vasodilator activity and anti-proliferative function has a great role on the wound healing process. [108] Sufficient amount of NO produced by endothelial cells, fibroblasts and leukocytes maintain the vascular tone and proliferation rate to restore the integrity of the injured skin. The following study aimed to show the difference in NOS expression and NO production between fibroblasts from normal skin and fibroblasts from HTS.

Skin samples were taken from six patients who suffered thermal injuries and were treated at Firefighter's Burn Treatment Unit and outpatient burn clinic, The University of Alberta Hospital and developed abnormal scars that resemble clinical features of HTS, after obtaining informed consents of them and site match normal skin sample were taken from uninjured skin from each patient as controls. These samples were obtained as 6 mm punch biopsy under local anesthesia. The fibroblasts from each tissue sample were isolated and cultured.

NO production was quantified using spectrophotometer and citrulline assay was used to determine Ca^{2+} dependent and independent NOS activity, RT-PCR was used to analyze mRNA expression of eNOS and iNOS in normal skin and HTS fibroblasts. It shows that both normal skin and HTS fibroblasts produce NO without any stimulation however HTS fibroblasts produce lower amounts of NO compared to normal skin ($p<0.01$, $n=6$, paired t-test). Fibroblasts from normal and HTS skin were stimulated with IFN-γ and LPS alone or in combination for 96 hours which reveled that there was no significant stimulation for NO production when IFN-γ and LPS used alone, but there was a significant increase in NO production from both normal and HTS fibroblasts when stimulated by combination of IFN-γ and LPS together and it was a concentration dependent response and the proportion of this increase was similar for both HTS and normal skin fibroblasts.

The mRNA for eNOS and iNOS were both expressed in normal and HTS fibroblasts and also stained positive on HTS and normal skin fibroblast however the HTS fibroblast showed lower staining intensity than the normal fibroblasts. In addition, the endothelial eNOS expression was decreased in HTS. It has been reported that the TGF-β suppress the NO production [109] and HTS express more TGF-β more than normal skin. [110] The above data suggest that a decrease of eNOS expression and NO production plays a role in increasing cellularity of the wound, leading to HTS formation.

Dermal Fibroblasts from HTS Produce Less Colleagenase [107]

HTS is pathological scar that is common clinical problem to patients who survives deep dermal injuries and it is a dermal form of fibroproliferative disorders, which characterized by accumulation of excess ECM proteins. Collagenase is a member of the matrix metalloproteinases family that starts the process of collagen degradation. [111]

During wound healing, collagenase has an important role in collagen deposition and connective tissue remodeling. [112] The production and its activity was examined Abergel et al. [113] which reported that collagenase activity in three out of six keloid fibroblasts were below the average of their normal controls. Insulin-like growth factor 1 (IGF-1) and TGF-β1 have been reported to be highly expressed in HTS in which they influence the collagen deposition and reduce collagenase activity. [23, 114]

Patients who suffered thermal injuries and developed clinical features that resembled HTS and were treated in University of Alberta Firefighters' Burn Treatment Unit were subjected to this study. Following an informed consent taken from the patients tissue samples were taken from HTS and site match normal skin by punch biopsy under local anesthesia.

The result showed the decreased mRNA expression of collagenase in HTS compared with normal skin. Interestingly, collagenase mRNA expression was reduce in both normal and HTS samples after treating with IGF-1; however the HTS fibroblasts were less responsive to IGF-1 which may be due to that fibroblasts from HTS already have lower expression for collagenase mRNA.

This inhibitory effect of IGF-1 may be not conclusive to explain the reduction in collagenase production in HTS and this warrant remains further investigation.

Fibroblasts from HTS Enhanced Expression of IGF-1 [110]

Fibroprolifirative disease is pathological disorders characterized by the disruption of ECM composition and organization. They involve many human tissue diseases such as pulmonary fibrosis, myelofibrosis, atherosclerosis, liver fibrosis and rheumatoid arthritis. [115] HTS is dermal form of these disorders, which usually follow extensive thermal skin injuries. [116]

However the development of HTS may be due to the interaction of many cells with cytokines and growth factors (e.g., IGF-1) and it is suggested that high level of IGF-1 may be the cause of type I and type III procollagen in post-burn HTS. [117]

The following study aimed to study the fibrogenic function of IGF-1 in dermal fibroblast and it role in HTS formation. Following informed consent taken from patients who suffered thermal injuries and were treated in University Of Alberta Hospital Firefighters' Burn treatment Unit who developed a red, hard, itchy scars that are confined to the scar boundaries skin biopsy were taken from these scar with site match normal skin as controls. Analysis using Northern blot shows that the mRNA expression for IGF-1 was markedly higher in HTS than the normal control skin in addition, treating fibroblast with IGF-1 caused increase in expression of type I and type III procollagen mRNA.

Thus the persistence of IGF-1 in post-burn HTS contributes in ECM accumulation, which is evident in all fibroproliferative diseases including HTS.

Conclusion

HTS is pathological scars that follows deep dermal or thermal injuries and characterized to be hyperemic, elevated, itchy scar that remain intact to its original borders and deposition of excess ECM deposition, resulting in HTS, a dermal form of fibroprolifirative disorders. The pathophysiology of HTS remains unclear, making HTS difficult to treat. Many cells and growth factors have been linked to HTS. Our previous data showed that fibroblasts from different layer of the skin had distinctive features in which the deep dermal fibroblasts proliferated in slower rate and produced more TGF-β1 and CTGF compared with superficial fibroblasts. We also showed that deep dermal fibroblasts produced more versican but less decorin and fibromdulin, more collagen but less collagenase compared with superficial fibroblast suggest that deep fibroblast may resemble HTS fibroblasts.

Moreover, the phenotype for fibroblasts appeared to be changed. The profile of molecular expression was found to be altered. Fibroblasts in HTS were more like myofibroblasts and resistant to apoptosis. After dermal injury, fibroblasts appeared to acquire immunoinflmamtory property. Better understanding the dermal fibroblast heterogeneity and their altered phenotypes in wound healing may give rise to a novel preventive and therapeutic modality for HTS.

References

[1] Tredget, E. E. Pathophysiology and treatment of fibroproliferative disorders following thermal injury. *Ann. N. Y. Acad. Sci.* 888, 165–182 (1999).

[2] Armour, A., Scott, P. G. and Tredget, E. E. Cellular and molecular pathology of HTS: basis for treatment. *Wound Repair Regen.* 15 Suppl. 1, S6–17 (2007).

[3] Sorrell, J. M., Baber, M. A. and Caplan, A. I. Site-matched papillary and reticular human dermal fibroblasts differ in their release of specific growth factors/cytokines and in their interaction with keratinocytes. *J. Cell. Physiol.* 200, 134–145 (2004).

[4] Reynolds, A. J., Oliver, R. F. and Jahoda, C. A. Dermal cell populations show variable competence in epidermal cell support: stimulatory effects of hair papilla cells. *Journal of Cell Science* 98 (Pt 1), 75–83 (2005).

[5] Harper, R. A. and Grove, G. Human skin fibroblasts derived from papillary and reticular dermis: differences in growth potential in vitro. *Science* 204, 526–527 (1979).

[6] Azzarone, B. and Macieira-Coelho, A. Heterogeneity of the kinetics of proliferation within human skin fibroblastic cell populations. *Journal of Cell Science* 57, 177–187 (1982).

[7] Schafer, I. A., Pandy, M., Ferguson, R., and Davis, B. R. Comparative observation of fibroblasts derived from the papillary and reticular dermis of infants and adults: growth kinetics, packing density at confluence and surface morphology. *Mech. Ageing Dev.* 31, 275–293 (1985).

[8] Reynolds, A. J. and Jahoda, C. A. Hair matrix germinative epidermal cells confer follicle-inducing capabilities on dermal sheath and high passage papilla cells. *Development* 122, 3085–3094 (1996).

[9] Jahoda, C. A. and Reynolds, A. J. Dermal-epidermal interactions. Adult follicle-derived cell populations and hair growth. *Dermatol. Clin.* 14, 573–583 (1996).

[10] Ali-Bahar, M., Bauer, B., Tredget, E. E., and Ghahary, A. Dermal fibroblasts from different layers of human skin are heterogeneous in expression of collagenase and types I and III procollagen mRNA. *Wound Repair Regen.* 12, 175–182 (2004).

[11] Sorrell, J. M., Baber, M. A. and Caplan, A. I. Clonal characterization of fibroblasts in the superficial layer of the adult human dermis. *Cell Tissue Res.* 327, 499–510 (2007).

[12] Ross, E. V., et al. Comparison of carbon dioxide laser, erbium:YAG laser, dermabrasion, and dermatome: a study of thermal damage, wound contraction, and wound healing in a live pig model: implications for skin resurfacing. *Journal of American Dermatology* 42, 92–105 (2000).

[13] Seukeran, D. C., Collins, P. and Sheehan-Dare, R. A. Adverse reactions following pulsed tunable dye laser treatment of port wine stains in 701 patients. *Br. J. Dermatol.* 136, 725–729 (1997).

[14] Cuttle, L., et al. A porcine deep dermal partial thickness burn model with hypertrophic scarring. *Burns* 32, 806–820 (2006).

[15] Dunkin, C. S., J. et al. Scarring Occurs at a Critical Depth of Skin Injury: Precise Measurement in a Graduated Dermal Scratch in Human Volunteers. *Plastic and Reconstructive Surgery* 119, 1722–1732 (2007).

[16] Wang, J., Dodd, C., Shankowsky, H. A., Scott, P. G., and Tredget, E. E. Deep dermal fibroblasts contribute to hypertrophic scarring. *Lab. Invest.* 88, 1278–1290 (2008).

[17] Honardoust, D., Ding, J., Varkey, M., Shankowsky, H. A., and Tredget, E. E. Deep Dermal Fibroblasts Refractory to Migration and Decorin-Induced Apoptosis Contribute to Hypertrophic Scarring. *Journal of Burn Care and Research* 33, 668–677 (2012).

[18] Honardoust, D., Varkey, M., Marcoux, Y., Shankowsky, H. A., and Tredget, E. E. Reduced Decorin, Fibromodulin, and Transforming Growth Factor-β3 in Deep Dermis Leads to Hypertrophic Scarring. *Journal of Burn Care and Research* 33, 218–227 (2012).

[19] Wang, J., et al. Increased TGF-beta-producing CD4+ T lymphocytes in postburn patients and their potential interaction with dermal fibroblasts in hypertrophic scarring. *Wound Repair Regen.* 15, 530–539 (2007).

[20] Yang, L., et al. Peripheral blood fibrocytes from burn patients: identification and quantification of fibrocytes in adherent cells cultured from peripheral blood mononuclear cells. *Lab. Invest.* 82, 1183–1192 (2002).

[21] Tredget, E. E., Yang, L., Delehanty, M., Shankowsky, H., and Scott, P. G. Polarized Th2 cytokine production in patients with hypertrophic scar following thermal injury. *J. Interferon Cytokine Res.* 26, 179–189 (2006).

[22] Prud'homme, G. J. and Piccirillo, C. A. The inhibitory effects of transforming growth factor-beta-1 (TGF-beta1) in autoimmune diseases. *J. Autoimmun.* 14, 23–42 (2000).

[23] Ghahary, A., Shen, Y. J., Scott, P. G., and Tredget, E. E. Expression of mRNA for transforming growth factor-beta 1 is reduced in hypertrophic scar and normal dermal fibroblasts following serial passage in vitro. *J. Invest. Dermatol.* 103, 684–686 (1994).

[24] Van Obberghen-Schilling, E., Roche, N. S., Flanders, K. C., Sporn, M. B., and Roberts, A. B. Transforming growth factor beta 1 positively regulates its own expression in normal and transformed cells. *J. Biol. Chem.* 263, 7741–7746 (1988).

[25] Wang, R., et al. Hypertrophic scar tissues and fibroblasts produce more transforming growth factor-beta1 mRNA and protein than normal skin and cells. *Wound Repair Regen.* 8, 128–137 (2000).

[26] Wang, J. F., Olson, M. E., Reno, C. R., Wright, J. B., and Hart, D. A. The pig as a model for excisional skin wound healing: characterization of the molecular and cellular biology, and bacteriology of the healing process. *Comp. Med.* 51, 341–348 (2001).

[27] Leask, A. and Abraham, D. J. The role of connective tissue growth factor, a multifunctional matricellular protein, in fibroblast biology. *Biochem. Cell Biol.* 81, 355–363 (2003).

[28] Wang, J. F., Olson, M. E., Ball, D. K., Brigstock, D. R., and Hart, D. A. Recombinant connective tissue growth factor modulates porcine skin fibroblast gene expression. *Wound Repair Regen.* 11, 220–229 (2003).

[29] Brigstock, D. R. The CCN family: a new stimulus package. *J. Endocrinol.* 178, 169–175 (2003).

[30] Colwell, A. S., Phan, T.-T., Kong, W., Longaker, M. T., and Lorenz, P. H. Hypertrophic scar fibroblasts have increased connective tissue growth factor expression after transforming growth factor-beta stimulation. *Plastic and Reconstructive Surgery* 116, 1387–90– discussion 1391–2 (2005).

[31] Hinz, B. Formation and Function of the Myofibroblast during Tissue Repair. *J. Investig. Dermatol.* 127, 526–537 (2007).

[32] Darby, I. A. and Hewitson, T. D. Fibroblast differentiation in wound healing and fibrosis. *Int. Rev. Cytol.* 257, 143–179 (2007).

[33] Desmoulière, A., Chaponnier, C. and Gabbiani, G. Tissue repair, contraction, and the myofibroblast. *Wound Repair and Regeneration* 13, 7–12 (2005).

[34] Pinnell, S. R. Regulation of collagen synthesis. *J. Invest. Dermatol.* 79 Suppl. 1, 73s–76s (1982).

[35] Ghahary, A., et al. Collagenase production is lower in post-burn hypertrophic scar fibroblasts than in normal fibroblasts and is reduced by insulin-like growth factor-1. *J. Invest. Dermatol.* 106, 476–481 (1996).

[36] Rudolph, R. Wide spread scars, hypertrophic scars, and keloids. *Clin. Plast. Surg.* 14, 253–260 (1987).

[37] Cohen, I. K. Can collagen metabolism be controlled: theoretical considerations. *J. Trauma* 25, 410–412 (1985).

[38] Diegelmann, R. F., Cohen, I. K. and McCoy, B. J. Growth kinetics and collagen synthesis of normal skin, normal scar and keloid fibroblasts in vitro. *J. Cell. Physiol.* 98, 341–346 (1979).

[39] Nagata, K. [Collagen-specific molecular chaperone, HSP47]. *Tanpakushitsu Kakusan Koso* 41, 888–895 (1996).

[40] Scott, P. G., Dodd, C. M., Tredget, E. E., Ghahary, A., and Rahemtulla, F. Chemical characterization and quantification of proteoglycans in human post-burn hypertrophic and mature scars. *Clin. Sci.* 90, 417–425 (1996).

[41] Varkey, M., Ding, J. and Tredget, E. E. Differential collagen–glycosaminoglycan matrix remodeling by superficial and deep dermal fibroblasts: Potential therapeutic targets for hypertrophic scar. *Biomaterials* 32, 7581–7591 (2011).

[42] McDougall, S., Dallon, J., Sherratt, J., and Maini, P. Fibroblast migration and collagen deposition during dermal wound healing: mathematical modelling and clinical implications. *Philosophical Transactions of the Royal Society A: Mathematical, Physical and Engineering Sciences* 364, 1385–1405 (2006).

[43] Nedelec, B., Shankowsky, H., Scott, P. G., Ghahary, A., and Tredget, E. E. Myofibroblasts and apoptosis in human hypertrophic scars: The effect of interferon-α2b. *Surgery* 130, 798–808 (2001).

[44] Wu, H., et al. Overexpression of decorin induces apoptosis and cell growth arrest in cultured rat mesangial cells in vitro. *Nephrology* 13, 607–615 (2008).

[45] Häkkinen, L., et al. A role for decorin in the structural organization of periodontal ligament. *Lab. Invest.* 80, 1869–1880 (2000).

[46] Hsieh, P. and Chen, L. B. Behavior of cells seeded in isolated fibronectin matrices. *J. Cell Biol.* 96, 1208–1217 (1983).

[47] Montgomery, R. B., Moscatello, D. K., Wong, A. J., Cooper, J. A., and Stahl, W. L. Differential modulation of mitogen-activated protein (MAP) kinase/extracellular signal-related kinase kinase and MAP kinase activities by a mutant epidermal growth factor receptor. *J. Biol. Chem.* 270, 30562–30566 (1995).

[48] Seidler, D. G., et al. Decorin protein core inhibits in vivo cancer growth and metabolism by hindering epidermal growth factor receptor function and triggering apoptosis via caspase-3 activation. *J. Biol. Chem.* 281, 26408–26418 (2006).

[49] Danielson, K. G., et al. Targeted disruption of decorin leads to abnormal collagen fibril morphology and skin fragility. *J. Cell Biol.* 136, 729–743 (1997).

[50] Scott, P. G., Dodd, C. M., Tredget, E. E., Ghahary, A., and Rahemtulla, F. Immunohistochemical localization of the proteoglycans decorin, biglycan and versican and transforming growth factor-beta in human post-burn hypertrophic and mature scars. *Histopathology* 26, 423–431 (1995).

[51] Zhang, Z., et al. Recombinant human decorin inhibits cell proliferation and downregulates TGF-beta1 production in hypertrophic scar fibroblasts. *Burns* 33, 634–641 (2007).

[52] Kalamajski, S. and Oldberg, A. The role of small leucine-rich proteoglycans in collagen fibrillogenesis. *Matrix Biol.* 29, 248–253 (2010).

[53] Hildebrand, A., et al. Interaction of the small interstitial proteoglycans biglycan, decorin and fibromodulin with transforming growth factor beta. *Biochem. J.* 302 (Pt 2), 527–534 (1994).

[54] Kolb, M., Margetts, P. J., Sime, P. J., and Gauldie, J. Proteoglycans decorin and biglycan differentially modulate TGF-beta-mediated fibrotic responses in the lung. *Am. J. Physiol. Lung Cell Mol. Physiol.* 280, L1327–34 (2001).

[55] Klass, B. R., Grobbelaar, A. O. and Rolfe, K. J. Transforming growth factor beta1 signalling, wound healing and repair: a multifunctional cytokine with clinical implications for wound repair, a delicate balance. *Postgrad. Med. J.* 85, 9–14 (2009).

[56] Cowin, A. J., Holmes, T. M., Brosnan, P., and Ferguson, M. W. Expression of TGF-beta and its receptors in murine fetal and adult dermal wounds. *Eur. J. Dermatol.* 11, 424–431 (2001).

[57] Roberts, A. B. and Sporn, M. B. Differential expression of the TGF-beta isoforms in embryogenesis suggests specific roles in developing and adult tissues. *Mol. Reprod. Dev.* 32, 91–98 (1992).

[58] Pohlers, D., et al. TGF-beta and fibrosis in different organs - molecular pathway imprints. *Biochim. Biophys. Acta* 1792, 746–756 (2009).

[59] Bandyopadhyay, B., et al. A 'traffic control' role for TGFbeta3: orchestrating dermal and epidermal cell motility during wound healing. *J. Cell Biol.* 172, 1093–1105 (2006).

[60] Colwell, A. S., Yun, R., Krummel, T. M., Longaker, M. T., and Lorenz, H. P. Keratinocytes modulate fetal and postnatal fibroblast transforming growth factor-beta and Smad expression in co-culture. *Plastic and Reconstructive Surgery* 119, 1440–1445 (2007).

[61] Goldberg, S. R., McKinstry, R. P., Sykes, V., and Lanning, D. A. Rapid closure of midgestational excisional wounds in a fetal mouse model is associated with altered transforming growth factor-beta isoform and receptor expression. *J. Pediatr. Surg.* 42, 966–71– discussion 971–3 (2007).

[62] Honardoust, D., Eslami, A., Larjava, H., and Häkkinen, L. Localization of small leucine-rich proteoglycans and transforming growth factor-beta in human oral mucosal wound healing. *Wound Repair Regen.* 16, 814–823 (2008).

[63] Hagood, J. S., Lasky, J. A., Nesbitt, J. E., and Segarini, P. Differential expression, surface binding, and response to connective tissue growth factor in lung fibroblast subpopulations. *Chest* 120, 64S–66S (2001).

[64] Zhou, Y., Hagood, J. S. and Murphy-Ullrich, J. E. Thy-1 expression regulates the ability of rat lung fibroblasts to activate transforming growth factor-beta in response to fibrogenic stimuli. *Am. J. Pathol.* 165, 659–669 (2004).

[65] Hagood, J. S., et al. Loss of fibroblast Thy-1 expression correlates with lung fibrogenesis. *Am. J. Pathol.* 167, 365–379 (2005).

[66] Varkey, M., Ding, J. and Tredget, E. E. Differential collagen-glycosaminoglycan matrix remodeling by superficial and deep dermal fibroblasts: Potential therapeutic targets for hypertrophic scar. *Biomaterials* 32, 7581–7591 (2011).

[67] Boyce, S. T., et al. Cultured skin substitutes reduce donor skin harvesting for closure of excised, full-thickness burns. *Ann. Surg.* 235, 269–279 (2002).

[68] Monstrey, S., Hoeksema, H., Verbelen, J., Pirayesh, A., and Blondeel, P. Assessment of burn depth and burn wound healing potential. *Burns* 34, 761–769 (2008).

[69] Sorrell, J. M. Fibroblast heterogeneity: more than skin deep. *Journal of Cell Science* 117, 667–675 (2004).

[70] Martin, P. Wound Healing--Aiming for Perfect Skin Regeneration. *Science* 276, 75–81 (1997).

[71] Singer, A. J. and Clark, R. A. Cutaneous wound healing. *N. Engl. J. Med.* 341, 738–746 (2000).

[72] Peterson, J. M., Barbul, A., Breslin, R. J., Wasserkrug, H. L., and Efron, G. Significance of T-lymphocytes in wound healing. *Surgery* 102, 300–305 (1987).

[73] Barbul, A., Shawe, T., Frankel, H. L., Efron, J. E., and Wasserkrug, H. L. Inhibition of wound repair by thymic hormones. *Surgery* 106, 373–6– discussion 376–7 (1989).

[74] Efron, J. E., Frankel, H. L., Lazarou, S. A., Wasserkrug, H. L., and Barbul, A. Wound healing and T-lymphocytes. *J. Surg. Res.* 48, 460–463 (1990).

[75] Davis, P. A., Corless, D. J., Aspinall, R., and Wastell, C. Effect of CD4(+) and CD8(+) cell depletion on wound healing. *Br. J. Surg.* 88, 298–304 (2001).

[76] Scott, S., Pandolfi, F. and Kurnick, J. T. Fibroblasts mediate T cell survival: a proposed mechanism for retention of primed T cells. *J. Exp. Med.* 172, 1873–1876 (1990).

[77] Geiler, T., Kriegsmann, J., Keyszer, G. M., Gay, R. E., and Gay, S. A new model for rheumatoid arthritis generated by engraftment of rheumatoid synovial tissue and normal human cartilage into SCID mice. *Arthritis Rheum.* 37, 1664–1671 (1994).

[78] Takayanagi, H., et al. Suppression of arthritic bone destruction by adenovirus-mediated csk gene transfer to synoviocytes and osteoclasts. *J. Clin. Invest.* 104, 137–146 (1999).

[79] Nedelec, B., Dodd, C. M., Scott, P. G., Ghahary, A., and Tredget, E. E. Effect of interferon-alpha2b on guinea pig wound closure and the expression of cytoskeletal proteins in vivo. *Wound Repair Regen.* 6, 202–212 (1998).

[80] Nedelec, B., Shankowsky, H. A. and Tredget, E. E. Rating the resolving hypertrophic scar: comparison of the Vancouver Scar Scale and scar volume. *J. Burn Care Rehabil.* 21, 205–212 (2000).

[81] Moulin, V. R., et al. Normal skin wound and hypertrophic scar myofibroblasts have differential responses to apoptotic inductors. *J. Cell. Physiol.* 198, 350–358 (2004).

[82] Wassermann, R. J., et al. Differential production of apoptosis-modulating proteins in patients with hypertrophic burn scar. *J. Surg. Res.* 75, 74–80 (1998).

[83] Hump Death and decoy receptors and p53-mediated apoptosis. [Leukemia. 2000] - *PubMed* - NCBI. 1–7 (2011).

[84] Linge, C., et al. Hypertrophic scar cells fail to undergo a form of apoptosis specific to contractile collagen-the role of tissue transglutaminase. *J. Invest. Dermatol.* 125, 72–82 (2005).

[85] Leask, A. and Abraham, D. J. TGF-beta signaling and the fibrotic response. *FASEB J.* 18, 816–827 (2004).

[86] Scott, P. G., Ghahary, A. and Tredget, E. E. Molecular and cellular aspects of fibrosis following thermal injury. *Hand Clin.* 16, 271–287 (2000).

[87] Tredget, E. E., Nedelec, B., Scott, P. G., and Ghahary, A. Hypertrophic scars, keloids, and contractures. The cellular and molecular basis for therapy. *Surg. Clin. North Am.* 77, 701–730 (1997).

[88] Xia, W., Phan, T.-T., Lim, I. J., Longaker, M. T., and Yang, G. P. Complex epithelial-mesenchymal interactions modulate transforming growth factor-beta expression in keloid-derived cells. *Wound Repair Regen.* 12, 546–556 (2004).

[89] Tredget, E. E., et al. Transforming growth factor-beta in thermally injured patients with hypertrophic scars: effects of interferon alpha-2b. *Plastic and Reconstructive Surgery* 102, 1317–28– discussion 1329–30 (1998).

[90] Chen, W., et al. Ontogeny of expression of transforming growth factor-beta and its receptors and their possible relationship with scarless healing in human fetal skin. *Wound Repair Regen.* 13, 68–75 (2005).

[91] Honardoust, D., et al. Small leucine-rich proteoglycans, decorin and fibromodulin, are reduced in postburn hypertrophic scar. *Wound Repair and Regeneration* 19, 368–378 (2011).

[92] Iozzo, R. V. The biology of the small leucine-rich proteoglycans. Functional network of interactive proteins. *J. Biol. Chem.* 274, 18843–18846 (1999).
[93] Soo, C., et al. Differential expression of fibromodulin, a transforming growth factor-beta modulator, in fetal skin development and scarless repair. *Am. J. Pathol.* 157, 423–433 (2000).
[94] Stoff, A., et al. Effect of adenoviral mediated overexpression of fibromodulin on human dermal fibroblasts and scar formation in full-thickness incisional wounds. *J. Mol. Med.* 85, 481–496 (2007).
[95] Zhang, Z., Liu, Y., Zhang, X., and Xu, W.-S. [The content of decorin and its mRNA expression in normal human skin and hyperplastic scars]. *Zhonghua Shao Shang Za Zhi* 20, 76–78 (2004).
[96] Wang, R., Ghahary, A., Shen, Y. J., Scott, P. G., and Tredget, E. E. Nitric oxide synthase expression and nitric oxide production are reduced in hypertrophic scar tissue and fibroblasts. *J. Invest. Dermatol.* 108, 438–444 (1997).
[97] Benyó, Z., Kiss, G., Szabó, C., Csáki, C., and Kovách, A. G. Importance of basal nitric oxide synthesis in regulation of myocardial blood flow. *Cardiovasc. Res.* 25, 700–703 (1991).
[98] Strapková, A., Antosová, M. and Nosálová, G. Relation of L-arginine to airway hyperreactivity. *Gen. Physiol. Biophys.* 27, 85–91 (2008).
[99] Takahashi, T. Pathophysiological significance of neuronal nitric oxide synthase in the gastrointestinal tract. *J. Gastroenterol.* 38, 421–430 (2003).
[100] Da Silva Krause, M., et al. Physiological concentrations of interleukin-6 directly promote insulin secretion, signal transduction, nitric oxide release, and redox status in a clonal pancreatic β-cell line and mouse islets. *J. Endocrinol.* 214, 301–311 (2012).
[101] Hibbs, J. B., Taintor, R. R., Vavrin, Z., and Rachlin, E. M. Nitric oxide: a cytotoxic activated macrophage effector molecule. *Biochem. Biophys. Res. Commun.* 157, 87–94 (1988).
[102] Snyder, S. H. Biosensors. NO release for good measure. *Nature* 358, 623 (1992).
[103] Schneemann, M., et al. Nitric oxide synthase is not a constituent of the antimicrobial armature of human mononuclear phagocytes. *J. Infect. Dis.* 167, 1358–1363 (1993).
[104] Marsden, P. A., et al. Molecular cloning and characterization of human endothelial nitric oxide synthase. *FEBS Lett.* 307, 287–293 (1992).
[105] Nakane, M., Schmidt, H. H., Pollock, J. S., Förstermann, U., and Murad, F. Cloned human brain nitric oxide synthase is highly expressed in skeletal muscle. *FEBS Lett.* 316, 175–180 (1993).
[106] Xie, Q. W., et al. Cloning and characterization of inducible nitric oxide synthase from mouse macrophages. *Science* 256, 225–228 (1992).
[107] Ghahary, A., et al. Collagenase production is lower in post-burn hypertrophic scar fibroblasts than in normal fibroblasts and is reduced by insulin-like growth factor-1. *J. Invest. Dermatol.* 106, 476–481 (1996).
[108] Jorens, P. G., Van Overveld, F. J., Vermeire, P. A., Bult, H., and Herman, A. G. Synergism between interleukin-1 beta and interferon-gamma, an inducer of nitric oxide synthase, in rat lung fibroblasts. *Eur. J. Pharmacol.* 224, 7–12 (1992).
[109] Vodovotz, Y., Bogdan, C., Paik, J., Xie, Q. W., and Nathan, C. Mechanisms of suppression of macrophage nitric oxide release by transforming growth factor beta. *J. Exp. Med.* 178, 605–613 (1993).

[110] Ghahary, A., Shen, Y. J., Nedelec, B., Scott, P. G., and Tredget, E. E. Enhanced expression of mRNA for insulin-like growth factor-1 in post-burn hypertrophic scar tissue and its fibrogenic role by dermal fibroblasts. *Mol. Cell. Biochem.* 148, 25–32 (1995).

[111] Porras-Reyes, B. H., Blair, H. C., Jeffrey, J. J., and Mustoe, T. A. Collagenase production at the border of granulation tissue in a healing wound: macrophage and mesenchymal collagenase production in vivo. Connect. *Tissue Res.* 27, 63–71 (1991).

[112] Mann, J. S., Kindy, M. S., Edwards, D. R., and Curry, T. E. Hormonal regulation of matrix metalloproteinase inhibitors in rat granulosa cells and ovaries. *Endocrinology* 128, 1825–1832 (1991).

[113] Abergel, R. P., et al. Biochemical composition of the connective tissue in keloids and analysis of collagen metabolism in keloid fibroblast cultures. *J. Invest. Dermatol.* 84, 384–390 (1985).

[114] Edwards, D. R., et al. Transforming growth factor beta modulates the expression of collagenase and metalloproteinase inhibitor. *EMBO J.* 6, 1899–1904 (1987).

[115] Bitterman, P. B. and Henke, C. A. Fibroproliferative disorders. *Chest* 99, 81S–84S (1991).

[116] Linares, H. A., Kischer, C. W., Dobrkovsky, M., and Larson, D. L. The histiotypic organization of the hypertrophic scar in humans. *J. Invest. Dermatol.* 59, 323–331 (1972).

[117] Ghahary, A., Shen, Y. J., Scott, P. G., and Tredget, E. E. Expression of fibronectin messenger RNA in hypertrophic and normal dermal tissues and in vitro regulation by interferon alfa-2b. *Wound Repair Regen.* 1, 166–174 (1993).

[118] Salmon, M., Scheel-Toellner, D., Huissoon, A. P., Pilling, D., Shamsadeen, N., Hyde, H., et al. Inhibition of T cell apoptosis in the rheumatoid synovium. *The Journal of Clinical Investigation*, 99(3), 439–446. doi:10.1172/JCI119178(1997).

[119] Gombert, W., Borthwick, N. J., Wallace, D. L., Hyde, H., Bofill, M., Pilling, D., et al. Fibroblasts prevent apoptosis of IL-2-deprived T cells without inducing proliferation: a selective effect on Bcl-XL expression. *Immunology*, 89(3), 397–404 (1996).

In: Dermal Fibroblasts
Editor: Xiaowen Bai
ISBN: 978-1-62618-799-3

Chapter 5

Role of Stromal Fibroblasts in Cancer Originated from Squamous Epithelia

Karel Smetana Jr.,[1,*] Barbora Dvořánková,[1] Pavol Szabo,[1] Hynek Strnad[2] and Michal Kolář[2]

[1] Charles University in Prague, 1st Faculty of Medicine, Institute of Anatomy, Prague, Czech Republic

[2] Institute of Molecular Genetics, Academy of Sciences of the Czech Republic, Prague, Czech Republic

Abstract

Dermal fibroblasts are essential for the function of normal adult squamous epithelium under physiological conditions and in the course of wound healing. They also play a substantial role during the development of epithelial appendages such as hairs and teeth.

In this chapter, we focus on characterization of the fibroblasts located in tumor stroma, with particular stress on the tumors originating from squamous epithelium. These fibroblasts represent an important component of the tumor microenvironment, which is able to influence the biological properties of the tumors including their local aggressiveness and metastatic potential.

They differ from the normal fibroblasts in their transcription profile, as revealed by microarray analyses, and in their crosstalk with epithelial cells. We propose therapeutic manipulation of this interaction as a target for a new type of anticancer adjuvant biological therapy.

Keywords: Epithelial-mesenchymal interaction, squamous cell carcinoma, basal cell carcinoma, cancer-associated fibroblast, cancer therapy

---

* E-mail: karel.smetana@lf1.cuni.cz.

Anatomical and Developmental Notes

Squamous epithelium covers external (skin) and internal surfaces (e.g., oral cavity, vocal cords, esophagus) that are under mechanic stress. This type of epithelium is characterized by the presence of proliferating cells in the basal layer. Suprabasal cells are postmitotic under physiological conditions and they are connected by specific intercellular contacts, which are highly resistant to mechanical force. These cells have characteristics of terminally differentiated elements, the most superficial of them are desquamated and in some locations they are keratinized (Plzák et al., 2005).

Similarly to other cell populations, this epithelium possesses a pool of stem cells. Monopotent stem cells are located in the basal layer (mucosa and epidermis) and multipotent stem cells are present in the bulge region of the outer root sheath of the hair follicle in the area between the orifice of sebaceous gland and insertion of arrector pilli muscle (Watt, 2002; Blanpain and Fuchs, 2006). The same location is also occupied by multipotent stem cells originated in the neural crest. These cells are characterized by the differentiation potential similar to embryonic neural crest cells (Sieber-Blum et al., 2004; Krejčí and Grim, 2010). The spatial coexistence of the two stem-cell types evokes hypotheses on its functional consequences. However, no experimental data are available. The basement membrane separates the epithelial structures from the sheets of connective tissue (dermis in the case of skin), which is rich in fibroblasts producing the extracellular matrix and cytokines as well as in blood/lymphatic capillaries responsible for epithelium nutrition (Kierszenbaum and Tres, 2012).

The squamous epithelium originates predominantly from the ectoderm (e.g., epidermis or oral mucosa) or from the endoderm (e.g., epithelium of vocal cords or larynx) (Sadler and Sadler, 2011). The fibroblasts of the dermis or of the subepithelial connective tissue component originate in the mesoderm except for the orofacial region, in which they originate from the so-called ectomesenchyme that originates in the neural crest (Gilbert, 2000). Although the epithelial and connective tissue components are separated by the basement membrane, their crosstalk is important as is particularly exampled by epidermis covering the palms and soles. The fibroblasts isolated from this tissue are able to induce expression of specific keratins also in keratinocytes prepared from other parts of the body (Yamaguchi et al., 1999). The development of epithelial appendages (e.g., hair follicles, teeth) underlines the importance of the mutual epithelial-mesenchymal interaction, as the communication between the two cell types (i.e., fibroblasts and the epithelium) is necessary for their correct formation and function (Salazar-Ciudad, 2008, You et al., 2008; Schneider et al., 2009).

Cancer Stem Cells and the Stroma

Malignant transformation is based on the alteration of genetic material by which proliferation, ability to die by apoptosis, cell adhesion and migration are seriously affected. The transformed cells forming the tumor are highly heterogeneous in their differentiation status. Animal experiments with cancer cell grafting have demonstrated that only a very limited proportion of the cancer cells are able to initiate tumor formation. These cells display a very low differentiation status.

Moreover, they are phenotypically and functionally similar to normal adult tissue stem cells and exhibit, e.g., low proliferation rate, resistance to anoikis and exclusion of xenobiotics. Based on these observations, the population of the tumor-initiating cells has been dubbed cancer stem cells (Motlík et al., 2007; Sell, 2010). Normal tissue stem cells need a very special yet poorly defined microenvironment, so called niche, which is necessary for the maintenance of their stem cell phenotype (Walker et al., 2009). It is highly probable that cancer stem cells need a similar environment (Cabarcas et al., 2011).

The tumor does not consist solely of cancer cells. A significant part of the tumor is constituted by its stroma. The stroma itself is composed of fibroblasts producing extracellular matrix (ECM), of infiltrating leukocytes and of newly formed capillaries responsible for cancer cell nutrition (Polyak et al., 2009). Given its supportive role, the tumor stroma is also a well-suited candidate for the niche. This permissive microenvironment can support the function of the cancer stem cells and stimulate their local aggressiveness and secondary spread through the organism during the process of lymphatic/organ metastasis (Plzák et al., 2010). From this point of view, the malignant tumor is similar to a complicated ecosystem, and tumor progression depends on a complex network of intercellular interactions and activities (Kareva, 2011). The complexity of the problem necessitates a complex interdisciplinary approach enabling better understanding of tumor biology at the molecular and cellular levels and representing an important step to personalized treatment of patients.

Cancer Stromal Fibroblasts: Introduction

It has been well documented in pathological textbooks that the tumor stroma is significantly different from the normal connective tissue. From the morphological point of view, there is a similarity between wound healing and stromal reaction in tumors (Dvorak, 1986). In the majority of tumors originated from the epithelial tissue, the fibroblasts of the stroma very frequently express α-smooth muscle actin (α-SMA) and are more similar to myfibroblasts (MFib) than to fibroblasts (Fib). These Fib are usually called cancer-associated fibroblasts (CAF) (De Wever et al., 2008). Even the Fib isolated from the squamous cell carcinoma (SCC) of the head and neck with no expression of α-SMA at the protein level exhibit significant upregulation of transcription of the gene encoding this protein (Strnad et al., 2010). Interestingly, MFib are also present in the wound bed, where they are responsible for the process of wound contraction (Hinz, 2007). In addition to this function, the wound MFib are responsible for extensive production of ECM molecules and their prolonged presence in the wound can lead to hypertrophic scarification (Goldberg et al., 2007). The origin of CAF has not been fully elucidated. They may develop from local Fib by paracrine activation of local cells from the cancer cells in their vicinity. However, their origin in local macrophages, pericytes or endothelial cells has also been described (Mueller et al., 2007; De Wever et al., 2008; Haviv et al., 2009). CAF may even be formed by the epithelial-mesenchymal transition of the cancer epithelium, although this hypothesis needs verification (Qiu et al., 2008). This process is well known for its importance during metastasis. It is highly probable that some fibroblasts originated from the cancer cells remain in the tumor site in the form of CAF. Therapeutic targeting of the tumor microenvironment, and in particular of the stromal Fib, seems to be a promising option (Kenny et al., 2007).

Cancer Stromal Fibroblasts: Production of Extracellular Matrix

The tumor stroma is very rich in ECM and in particular in fibronectin, tenascin-C and galectin-1 (Gal-1) (Lacina et al., 2007a; Brellier and Chiquet-Ehrismann, 2012; Valach et al., 2012). This observation again finds its parallel in the process of wound healing, where all described molecules are also expressed (Grinell, 1984; Klíma et al., 2009; Midwood and Orend, 2009; Gál et al., 2011). Because the extracellular matrix molecules are also cleaved by proteases secreted by cancer cells, stromal fibroblasts and infiltrating leukocytes, the high expression of the ECM in the tumor stroma brings evidence about their rapid turnover in site. Gal-1 stimulates tumor growth (Szebeni et al., 2012; Valach et al., 2012). Production of both fibronectin and Gal-1 is also stimulated in normal dermal Fib by addition of TGF-β1 or Gal-1 separately or simultaneously (Brenmoehl et al., 2009; Dvořánková et al., 2011). Interestingly, after stimulation the normal dermal Fib are transformed to MFib exhibiting SMA and resemble CAF or MFib of the wound. The scaffold created by stimulated Fib contains both fibronectin and Gal-1, has the character of a complicated network and is suitable for cultivation of epithelial cells, e.g., keratinocytes. The keratinocytes cultured on this scaffold exhibit the differentiation pattern typical of the cells at the beginning of the differentiation cascade (Dvořánková et al., 2011). These observations suggest that the stromal reaction of CAF is not particular to tumors but is also important in reparative processes, as demonstrated by the control of wound healing in experiments in a rat model (Dvořánková et al., 2011).

Cancer Stromal Fibroblasts: Biological Activity

As observed by many authors for different types of malignant tumors, the stromal Fib significantly influence the biological properties of cancer cells, making them locally aggressive and forming lymphatic or distant metastases (Plzák et al., 2010). The intercellular contacts between cancer cells and Fib and their influence on the biology of cancer have been described (Li et al., 2003). Moreover, the majority of data demonstrate paracrine mechanisms of their crosstalk. In this section, we will focus on this paracrine signaling and describe several experiments documenting the crosstalk.

CAF isolated from the tumor. Fib isolated from basal cell carcinoma (BCCF) cocultured with normal human keratinocytes stimulate expression of keratin 19 and Ber-EP4 in originally negative cells (Lacina et al., 2007a). Similarly, the stromal Fib isolated from squamous cell carcinoma (SCCF) are significantly bioactive to normal epithelial cells when both cell types are cocultured. After coculture with SCCF, the normal keratinocytes express keratin 8, keratin 19 and in some cases also vimentin. The expression of keratin 8 seems to be very important, as this type of intermediate filament is not present in the normal postnatal squamous epithelia. Its presence is indeed a marker of poor prognosis for the patient (Gires et al., 2006). Cells expressing both pankeratin and vimentin frequently exhibit the signal for transcription factor Snail in their nuclei and can be considered as cells in the process of epithelial-mesenchymal transition. Because the medium conditioned by SCCF is also bioactive for keratinocytes, the paracrine effect of CAF on normal/malignant epithelium may be hypothesized. Transcriptionally, SCCF differ from the normal Fib in more than 500 genes.

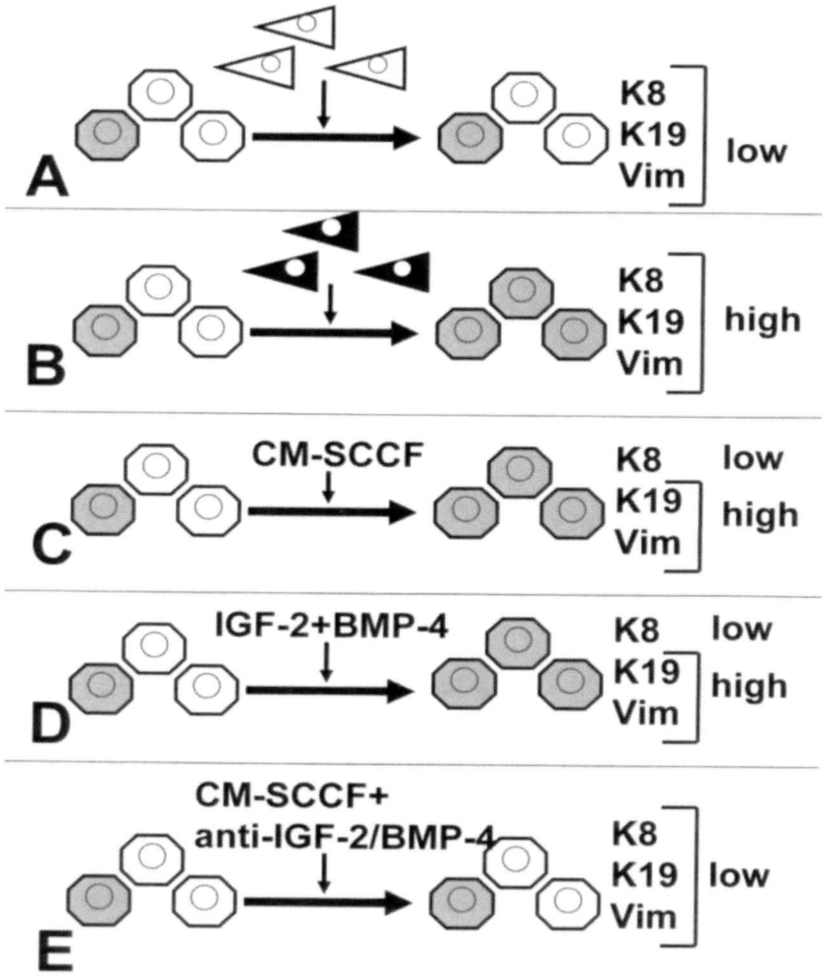

Figure 1. Only a very few normal human keratinocytes (hexagons) exhibit keratin 8 (K8), keratin 19 (K19) and both keratins and vimentin (Vim) *in vitro* even when they are cocultured with normal fibroblasts (triangles) (A). The fibroblasts isolated from the stroma of squamous cell carcinoma (black triangles) induce expression of the aforementioned markers in the keratinocytes (B). Conditioned medium, in which the fibroblasts isolated from squamous cell carcinoma have been cultured (CM-SCCF), stimulates expression of both keratin 19 and vimentin, but not of keratin 8 (C). The same effect as with the conditioned medium is also achieved by introduction of IGF-2 and BMP-4 to the cultivation system (D). Antibodies against the two growth factors are able to normalize the phenotype of keratinocytes cultured in the previously described conditioned medium (E).

Among them, growth factors IGF-2 and BMP-4 seem to contribute to the effects of stromal fibroblasts on the epithelium (Figure 1) (Lacina et al., 2007b; Strnad et al., 2010). The crucial role of TGF-β has also been established, as the genetic alteration of TGFBR1 is connected with spontaneous healing of multiple SCC of the skin (Goudie et al., 2011).

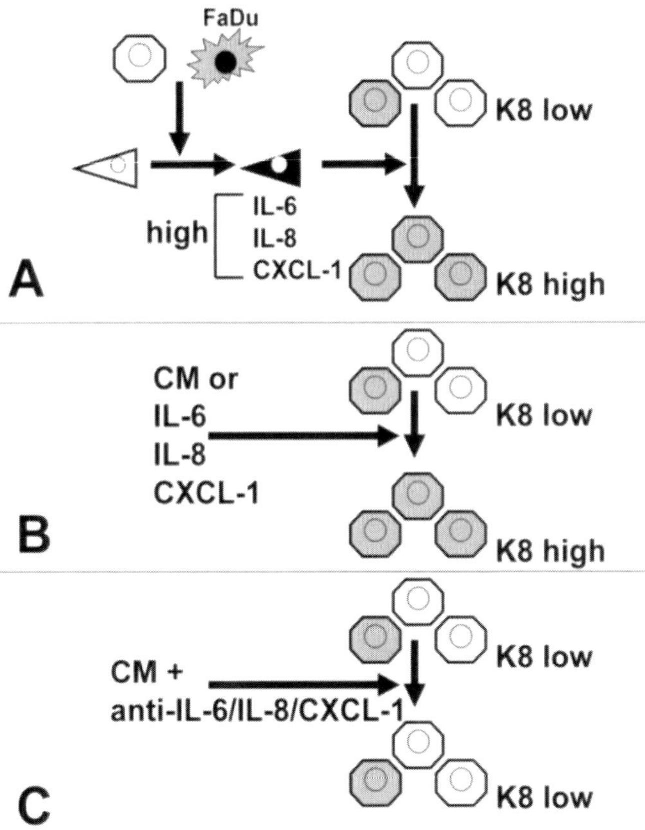

Figure 2. Normal keratinocytes (hexagons) or FaDu cancer cells influence the phenotype of normal fibroblasts (white triangles) in which production of IL-6, IL-8 and CXCL-1 is upregulated (black traingles). These activated fibroblasts strongly stimulate expression of keratin 8 (K8) in keratinocytes (A). The same effect is exhibited by the medium conditioned (CM) by coculture of the fibroblasts with FaDu cells (or fibroblasts from squamous cell carcinoma with FaDu cells) or the medium containing IL-6, IL-8 and CXCL-1 (B). The conditioned medium with blocked activity of the aforementioned factors has no effect on the keratinocytes (C).

Normal Fib induced by cancer cells. Because the local Fib activated by cancer cells represent one of the possible sources of CAF, we examined, with positive results, whether the FaDu cells prepared from SCC are able to influence the transcriptome of normal Fib so as to resemble the transcription profile of SCC. The activity of *IL-6*, *IL-8* and *CXCL-1* is crucial, as the three genes are significantly upregulated both when comparing SCC to the normal tissue and FaDu-influenced Fib to naive Fib. Fib precultured with FaDu cells display strong effects on normal keratinocytes, where they induce expression of keratin 8. The same effect can be induced by addition of IL-6, IL-8 and CXCl-1 to the cultivation medium and eliminated by addition of blocking antibodies to the conditioned medium (Figure 2). It should be mentioned here that the same activity as cancer epithelial cells is also displayed by normal keratinocytes, which are able to induce the above-described phenotype in normal Fib. This observation indicates the similarity between cancerogenesis and healing mentioned above (Smetana et al., 2012; Kolář et al., 2012).

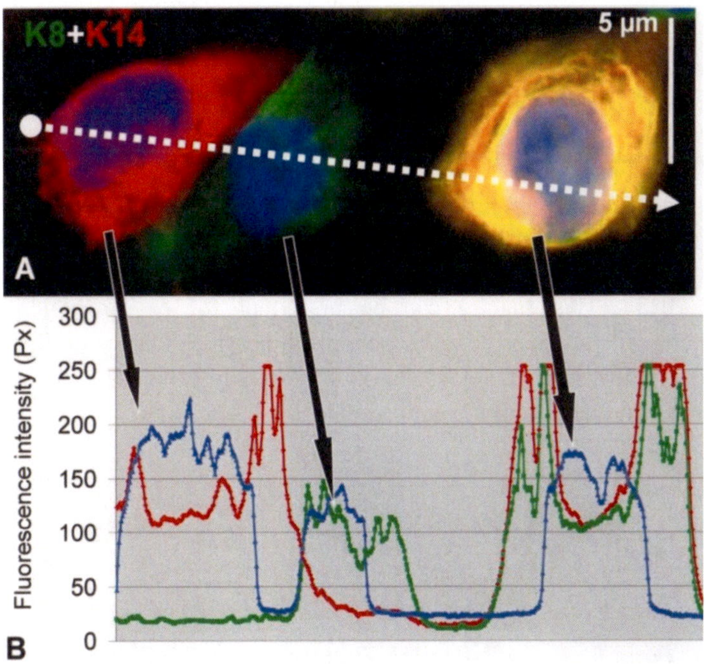

Figure 3. Culture of EM-G3 cells contains cells exhibiting a luminal marker (keratin 8, K8, green signal) and a basal myoepithelial marker (keratin 14, K14, red signal). The cells positive for both markers are typical in cultures of EM-G3 cells together with fibroblasts prepared from tumors, in this case from malignant melanoma (A). The fluorescence profiles demonstrate the colocalization of the signals of the two keratins (B).

Fib-like elements created by epithelial-mesenchymal transition of cancer cells. The cancer epithelial cells after epithelial-mesenchymal transition are also hypothesized as a possible source of CAF. The fibroblasts formed in the cancer epithelium display the same alteration of genetic material as their precursors, i.e., cancer cells. In an experiment, mouse lung epithelial cells were transfected by HPV-16 genes *E6/7*. The transformed cells were of the typical Fib phenotype and they were highly oncogenic. When employed as a model, CAF formed from cancer cells induced the cocultured epithelial cells to significantly change their morphology (very small size) and to strongly express keratins 8 and 19 and, in the majority of elements, coexpress both keratins and vimentin (Smetana et al., 2008). This observation suggests that cancer-epithelium-originated fibroblast-like cells are able to influence the phenotype of cancer cells.

Universal means of the crosstalk. In many types of carcinomas, we see a similar effect of stromal Fib on cancer cells. The question then is whether the effect of CAF is tumor type specific. We have observed that various stromal Fib, isolated from either squamous cell carcinoma, basal cell carcinoma, skin metastases of breast carcinoma or skin metastases of melanoma, induce simultaneous expression of both keratin 14 and keratin 8 in the EM-G3 breast cancer cells. Since keratin 14 is a marker of basal myoepithelial cells and keratin 8 is a marker of luminal cells, these double-positive cells can be considered for precursor cells with properties close to stem cells. Their presence in clinical samples indeed signals very poor prognosis in cancer-suffering patients (Figure 3) (Dvořánková et al., 2012). This observation indicates that the effect of CAF on epithelial cells has a nonspecific character.

These results clearly indicate the effect of CAF on epithelial cells. However, practically nothing is known about the possible effect of CAF on normal Fib. Coculture experiments demonstrate that CAF are able to significantly influence the phenotype of prenatal mouse Fib. They are able to induce expression of pluripotency markers such as Oct-4, Nanog and Sox-2. Moreover, the induced cells possess the differentiation potential similar to mesenchymal stem cells. Analyzing the transcription differences between normal Fib and Fib isolated from BCC stroma, the genes encoding growth factors such as *IGF-2*, *FGF-7*, *LEP*, *NGF* and *TGF-β* are strongly upregulated and they can be behind the influence of CAF on the Fib (Figure 4) (Szabo et al., 2011). This finding can be important for cancer biology because the mesenchymal stem cells are considered as one of the possible sources of CAF (De Boeck et al., 2010). Moreover, the mesenchymal stem cells are able to stimulate both tumor growth and metastasis in animal models (Karnoub et al., 2007).

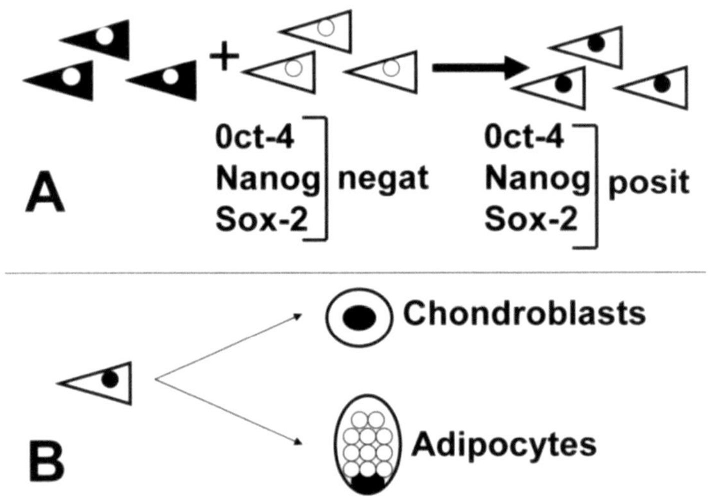

Figure 4. Fibroblasts isolated from basal cell carcinoma induce expression of markers of pluripotence such as Oct-4, Nanog and Sox-2 in cocultured mouse fetal fibroblasts (A). These fibroblasts can be differentiated to chondroblasts and adipocytes (B).

Conclusion and Perspectives

Concerning their origin, Fib of the tumor stroma represent a heterogeneous group. Disregarding the obscurity of their nature, they play a significant role in tumor growth and metastatic spread, probably by participating in the formation of the cancer-stem-cell niche and by supporting cancer cell proliferation and migration.

The mutual similarities between wound healing and cancerogenesis indicate a likely involvement of the epithelial-mesenchymal interaction in these processes. The presented data suggest that manipulations of the interaction between cancer stromal Fib and malignant epithelial cells have therapeutic potential as a new modality for anticancer therapy. Similarly, the findings may be utilized in the wound management. However, these phenomena warrant further investigation.

Acknowledgments

Part of the results published in this chapter have been supported by the Internal Grant Agency of the Ministry of Health of the Czech Republic, project No. NT 13488-4/2012, by Grant Agency of the Czech Republic, project No. 304/12/1333, and by Charles University, project PRVOUK No. 27-1.

References

Blanpain, C., Fuchs, E. (2006). Epidermal stem cells of the skin. *Annual Review of Cell and Developmental Biology 22*, 339-373.

Brellier, F., Chiquet-Ehrismann, R. (2012). How do tenascins influence the birth and life malignant cell? *Journal of Cellular and Molecular Medicine 16*, 32-40.

Brenmoehl, J., Miller, S. N., Hofmann, C., Vogl, D., Falk, W., Schölmerich, W., Rogler, G. (2009). Transforming growth factor-β1 induces intestinal myofibroblasts differentiation and modulates their migration. *World Journal of Gasgtroenterology 28*: 1431-1442.

Cabarcas, S. M., Mathews, L. A., Farrar, W. L. (2011). Cancer stem cell niche. *International Journal of Cancer* 129, 2315-2327.

De Boeck, A., Narine, K., De Neve, W., Mareel, M., Bracke, M., Se Wever, O. (2010). Resident and bone marrow-derived mesenchymal stem cells in head and neck squamous cell carcinoma. *Oral Oncology 45*, 336-342.

De Wever, O., Demetter, P., Mareel, M., Bracke. M. (2008). Stromal myofibroblasts are drivers of invasive cancer growth. *International Journal of Cancer 123*, 2229-2238.

Dvorak, H. F. (1986) Tumors: wounds that do not heal. Similarities between tumor stroma generation and wound healing. *New England Journal of Medicine 315*: 1650-1659.

Dvořánková, B., Szabo, P., Lacina, L., Gal, P., Uhrova J., Zima, T., Kaltner, H., André, S., Gabius, H.-J., Syková, E., Smetana, K. Jr. (2011) Human galectins induce conversion of dermal fibroblasts into myofibroblasts and production of extracellular matrix: potential application in tissue engineering and wound repair. *Cells Tissues Organs 194*, 469-480.

Dvořánková, B., Szabo, P., Lacina, L., Kodet, O., Matoušková, E., Smetana, K. Jr. (2012) Fibroblasts prepared from different types of malignant tumors stimulate expression of luminal marker keratin 8 in the EM-G3 breast cancer cell line. *Histochemistry and Cell Biology 137*, 679-685.

Gál, P., Vasilenko, T., Kostelníková, M., Jakubco, J., Kovác, I., Sabol, F., André, S., Kaltner, H., Gabius, H.-J., Smetana K., Jr. (2011). Open wound healing *in vivo*: monitoring, binding and presence of adhesion/growth-regulatory galectins in rat skin during the course of complete re-epithelialization. *Acta Histochemica et Cytochemica 44*, 191–199.

Gilbert S. F. (2000). Developmental Biology. Sinauer, Sunderland.

Gires, O., Mack, B., Rauch, J., Matthias, C. (2006). CK 8 correlates with malignancy in leukoplakia and carcinomas of the head and neck.. *Biochemical and Biophysical Research Communications 304*, 252-259.

Goldberg, M. T., Han, Y.-P., Yan, C., Shaw, M. C., Garner, W. L. (2007). TNF-α suppress α-smooth muscle actin expression in human dermal fibroblasts: An implification for abnormal wound healing. *Journal of Investigative Dermatology 127*: 2645-2655.

Goudie, D.R., D'Alessandro, M., Merriman, B., Lee, H., Szeverényi, I., Avery, S., O'Connor, B. D., Nelson, S. F., Coats, S. E., Stewart, A., Christie, L., Pichert, G., Friedel, J., Hayes, I., Burrows, N., Whittaker, S., Gerdes, A. M., Broesby-Olsen, S., Ferguson-Smith, M. A., Verma, C., Lunny, D. P., Reversade, B., Lane, E. B. (2011). Multiple self-healing squamous epithelioma is caused by a disease-specific spectrum of mutations in TGFBR1. *Nature Genetics 43*, 365-369.

Grinnell F. (1984). Fibronectin and wound healing. *Journal of Cellular Biochemistry* 26(2), 107-16.

Haviv, I., Polyak, K., Qiu, W., Hu, M., Campbell, I. (2009). Origin of cancer associated fibroblasts. *Cell Cycle 8*, 589-595.

Hinz, B. (2007). Formation and function of myofibroblasts during tissue repair. *Journal of investigative dermatology. 127*, 526-537.

Kareva, I. (2011). What can ecology teach us about cancer? *Translational Oncology 4*, 266-270.

Karnoub, A. E., Dash, A. B., Vo, A. P., Sullivan, A., Brooks, M. W., Bell, G. V. Richardson, A. L., Polyak, K., Tubo, R., Weinberg, R. A. (2007). Mesenchymal stem cells within tumour stroma promote breast cancer metastasis. *Nature 449*, 557-565.

Kenny, P. A., Lee, G. Y. Bissell, M. J. (2007). Targeting the tumor microenvironment. *Frontiers in Bioscience 12*. 3468-3474.

Kierszenbaum A., L., Tres, L., L. (2012). Histology and Cell Biology. *An Introduction to Pathology*. Elsevier, Philadelphia.

Klíma, J., Lacina, L., Dvořánková, B., Herrmann, D., Carnwath, J. W., Niemann, H., Kaltner, H., André, S., Motlík, J., Gabius, H.-J., Smetana, K. Jr. (2009). Differential regulation of galectin expression/reactivity during wound healing in porcine skin and in cultures of epidermal cells with functional impact on migration. *Physiological Research 58*, 873-884.

Kolář, M., Szabo, P., Dvořánková, B., Lacina, L., Gabius, H.-J., Strnad, H., Šáchová, J., Vlček, Č., Plzák, J., Chovanec, M., Čada, Z., Betka, J., Fík, Z., Pačes, J., Kovářová, H., Motlík, J., Jarkovská, K., Smetana, K. Jr. (2012) Upregulation of IL-6, IL-8 and CXCL-1 production in dermal fibroblasts by normal/malignant epithelial cells *in vitro*, immunohistochemical and transcriptomic analyse. *Biology of the Cell 104*, 738-751.

Krejčí, E., Grim, M. (2010). Isolation and characterization of neural crest stem cells from adult hair follicles. *Folia Biologica 56*, 149-157.

Lacina, L., Smetana, K. Jr., Dvořánková, B., Pytlík, R., Kideryová, L., Kučerová, L., Plzáková, Z., Štork, J., Gabius, H.-J., André, S. (2007a). Stromal fibroblasts from basal cell carcinoma affect phenotype of normal keratinocytes, British Journal of Dermatology *156*, 819-829.

Lacina, L., Dvořánkova, B., Smetana, K. Jr., Chovanec, M., Plzák, J., Tachezy, R., Kideryová, L, Kučerová, L., Čada, Z., Bouček, J., Kodet, R., André, S., Gabius, H.-J. (2007b). Marker profiling of normal keratinocytes identifies the stroma from squamous cell carcinoma of the oral cavity as a modulatory microenvironment in co-culture. *International Journal of Radiation Biology 83*, 837-848.

Li, G., Satyamoorthy, K., Meier, F., Berking, C., Bogenreeder, T., Herlyn, M. (2003). Function and regulation of melanoma-stromal fibroblast interactions: when seeds meet soil. *Oncogene 22*, 3162-3171.

Midwood, K. S., Orend, G. (2009). The role of tenascin-C in tissue injury and tumorigenesis. *Journal of Cell Communication and Signaling 3*, 287-310.

Motlík, J., Klíma, J., Dvořánková, B., Smetana, K. Jr. (2007). Porcine epidermal stem cells as a biomedical model for wound healing and normal/malignant epithelial cell propagation. *Theriogenology 67,* 105-111.

Mueller, L., Goumas, F. A., Affeldt, M., Sandtner, S., Gehling, U. M., Brilloff, S., Walter, J., Karnatz, N., Lamszus, K., Rogiers, S., Broering, D. C. (2007). Stromal fibroblasts in colorectal liver metastases originate from from resident fibroblasts and generate an inflammatory microenvironment, *American Journal of Pathology 171*, 1608-1618.

Plzák, J., Smetana, K., Jr., Chovanec, M., Betka, J. (2005). Glycobiology of head and neck squamous cell carcinoma. *Otorhinolaryngology 67*, 61-69.

Plzák, J., Lacina, L., Smetana, K. Jr., Chovanec, M., Dvořánková, B., Szabo, P., Čada, Z. (2010). Epithelial-stromal interaction in squamous cell epithelium-derived tunors : an important new player in the control of tumor biological properties. *Anticancer Research 30*, 455-462.

Polyak, K., Haviv, I., Campbell, I. G. (2009). Co-evolution of tumor cells and their microenvironment. *Trends in Genetics 25*, 30-38.

Qiu, W., Hu, M., Sridhar, A., Opeskin, K., Fox, S., Shipitsin, M., Trivett, M., Thompson, E. R., Ramakrishna, M., Gorringe, K. L., Polyak, K., Haviv, I., Campbell, I. G. (2008) No evidence of clonal somatic genetic alterations in cancer-associated fibroblasts from human breast and ovarian carcinomas. *Nature Genetics* 40, 650-655.

Sadler, T. W and Sadler T (2011). *Langemann´s Medical Embryology*. Lippincot Williams and Wilkins, Philadelphia.

Salazar-Ciudad, I. T. (2008). Tooth morphogenesis *in vivo*, *in vitro*, and *in silico*. *Current Topics in Developmental Biology 81*, 341-371.

Schneider, M. R., Schmidt-Ullrich, R., Paus, R. (2009). The hair follicle as a dynamic miniorgan. *Current Biology 19*, R132-142.

Sell, S. (2010). On the stem cell origin of cancer. *American Journal of Pathology 176*, 2584-2594.

Sieber-Blum, M., Grim, M., Hu, Y. F., Szeder, V. (2004). Pluripotent neural crest stem cells in the adult hair follicle. *Developmental Dynamics 231*, 258-269.

Smetana, K. Jr., Dvořánková, B., Lacina, L., Čada, L., Vonka, V. (2008). Human hair follicle and interfollicular keratinocyte reaktivity to mouse HPV16-transformed cells: An *in vitro* study. *Oncological Report 20*, 75-80.

Smetana, K. Jr., Dvořánková, B., Lacina, L., Strnad H., Kolář, M., Chovanec, M., Plzák J., Čada, Z., Vlček, Č., Szabo, P., Betka J., Motlík, J., Kovářová, H., Jarkovská, K. (2012). Combination of antibodies or their Fab fragments for employments as drug or pharmaceutic agent containing these antibodies or their Fab fragments. *Czech Patent No.* 303 227.

Strnad, H., Lacina, L., Kolář, M., Čada, Z., Vlček, Č., Dvořánková, B., Betka, J., Plzák, J., Chovanec, M., Šáchová, J., Valach, J., Urbanová, M., Smetana, K., Jr. (2010). Head and neck squamous cancer fibroblasts produce growth factors influencing phenotype of normal human keratinocytes. *Histochemistry and Cell Biology 133*, 201-211.

Szebeni, G. J., Kriston-Pál, E., Blazsó, P., Novák, J., Szabó, E., Czibula, Á., Fajka-Boja, R., Hegyi, B., Uher, F., Krenács, L., Joó, G., Monostori, E. (2012). Identification of galectin-

1 as a critical factor in function of mouse mesenchymal stromal cell- mediated tumor promotion. *PLOS One 7*, e41372.

Szabo, P., Kolář, M., Dvořánková, B., Lacina, L., Štork, J., Vlček, Č., Strnad, H., Tvrdek, M., Smetana, K. Jr. (2011). Mouse 3T3 fibroblasts under the influence of fibroblasts isolated from stroma of human basal cell carcinoma acquire properties of multipotent stem cells. *Biology of the Cell 103*, 233-248.

Valach, J., Fík, Z., Strnad, H., Chovanec, M., Plzák, J., Čada, Z., Szabó, P., Šáchová, J., Hroudová, M., Urbanová, M., Šteffl, M., Pačes, J., Mazánek, J., Vlček, Č., Betka, J., Kaltner, H., André, S., Gabius, H.-J., Kodet, R., Smetana, K. Jr., Gál, P., Kolář, M. (2012). Smooth muscle actin-expressing stromal fibroblasts in head and neck squamous cell carcinoma: increased expression of galectin-1 and induction of poor-prognosis factors. *Int. J. Cancer, 131*, 2499-2508, doi: 10.1002/ijc.27550.

Walker, M. R., Patel K. K., Stappenbeck, T. S. (2009). The stem cell niche. *Journal of Pathology 217*, 169-180.

Watt, F. M. (2002). Role of integrins in regulating of epidermal adhesion, growth and differentiation. *The EMBO Journal 21*, 3919-3926.

Yamaguchi, Y., Itami, S., Tarutani, M., Hosokawa, K., Miura, H., Yoshikawa, K. (1999). Regulation of keratin 9 in nonpalmoplantar keratinocytes by palmoplantar fibroblasts through epithelial-mesenchymal interactions. *Journal of Investigative Dermatology 112*, 483-488.

Yu, B. D., Mukhopadhyay, A., Wong , C. (2008). Skin and hair: models for exploring organ regeneration. *Human Molecular Genetics ,1*, R54-59.

In: Dermal Fibroblasts
Editor: Xiaowen Bai
ISBN: 978-1-62618-799-3

Chapter 6

The Role of Fibroblasts in Cancer Prone Genodermatoses

Yannick Gache and Thierry Magnaldo*
Genetics and physiopathology of epithelial cancers,
Institute For Research on Cancer and Aging, Nice / IRCAN,
School of medicine, Nice University

Abstract

The invasiveness of epithelial cancers arises with age and is known to depend on interactions between the tumor cells and their microenvironment. Precising the exact nature of those interactions deserves further investigations. Human skin is an excellent model for studying genetic and environmental factors contributing to carcinoma development upon aging. Our studies on cells isolated from patients presenting high susceptibilty to cancer, such as Xeroderma pigmentosum, or the Nevoid Basal Cell Carcinoma or Gorlin' syndrome, revealed that both epithelial and mesenchymal cells were affected in the absence of genotoxic stress and external stimuli. We modelized cancer-like features at the cutaneous level by developing organotypic cultures using the skin cells of patients. Notably, patients' fibroblasts overexpressed factors associated with extracellular matrix modifications (MMP1, MMP3), cell proliferation (KGF/FGF7), or inflammation (SDF1α/CXCL12), all hallmarks of dermal fibroblasts activated in the vicinity of epithelial tumors (Carcinoma-Associated Fibroblasts). Patients' fibroblasts could thus play a key role in early tumor development and aggressiveness. Deciphering the contribution of the stroma to the evolution of epithelial cancer is expected to improve antitumoral treatments.

* Corresponding Author address: Genetics and physiopathology of epithelial cancers, UMR CNRS 7284 INSERM U1081, Institute For Research on Cancer and Aging, Nice / IRCAN, School of medicine, Nice University, 28, Avenue de Valombrose 06107 Nice cedex 02, tel (33) 4 93 37 76 70, tel sec (33) 4 93 37 77 86, Fax : 04 93 37 70 92, GSM (33) 6 79 80 47 90, http://www.ircan.org/, Email: Thierry.Magnaldo@unice.fr.

Keywords: Skin, cancer, microenvironment, xeroderma pigmentosum, Nevoid Basal Cell Carcinoma/Gorlin Goltz syndrome

Introduction

Essential biological processes, such as embryonic development, tissular maintenance and repair, rely on highly controlled interactions beetwen epithelial cells and their complex microenvironment made of endothelial, immune, adipose, and fibroblast cells, producing growth factors and cytokines as well as macromolecules of the extracellular matrix. In the last decades, compelling evidence has accumulated indicating the essential role of mesenchymal fibroblasts in the fate of cancer cells (Polyak et al., 2009). Indeed, the malignant transformation of the epithelial cell is accompanied by severe qualitative and quantitative changes in epithelial-mesenchymal interactions that may favor tumor cell invasion and agressivity, promote genetic instability, and pull tumoral cells toward metastatic colonization of distal organs. Pathological conditions include prostate, lung, pancreas and skin cancers, all implying complex neo interactions beetwen the activated stroma and cancer epithelial cells. As a matter of public health concern, epithelial cancers make up about 80% of human cancers.

Skin Cancers

Due to its location on the body's surface, skin is a preferential target of genotoxic/mutagenic stresses such as ultraviolet radiations (UV). Skin is schematically composed of two distinct compartments, the epidermis and the dermis. The epidermis mainly comprises keratinocytes that are epithelial cells from an ectodermal origin. The epidermis also contains Merkell (neuro endocrine cells), Langerhans (dendritic immune cells) and melanocytes (cells responsible for the skin pigmentation that derive and migrate from the neural crest upon embryonic development). The dermis is the underlying compartment in which fibroblasts, from a mesenchymal origin, are the main cellular components. Fibroblasts are responsible for the synthesis of elastin and collagen fibers that make a physical support to the epidermis. At the steady state, dermal fibroblasts also secrete growth factors and cytokines that fuel the soil in a manner necessary for constant epidermal renewal. In case of skin injuries such as burns or cuts, dermal fibroblasts may also mount responses adapted to the situation, including secretion of growth factors, cytokines, extracellular matrix (ECM) components and matrix remodeling enzymes.

Non-melanoma skin cancers such as basal and squamous cell carcinomas (BCC and SCC) originate from keratinocytes and account for about 30% of human neoplasia. Malignant melanoma (MM), derived from melanocytes, account for about 2% of human cancers. Both SCC and MM are distinguished from BCC by their significant proneness to metastatic colonization. As a consequence, both SSC and MM are generally lethal when care occurs too late. For instance, according to the Breslow index, it is considered that the median life expectancy is about 5 years in the case of MM lesions with a deepness of more than 5 mm. Conversely, although BCCs are generally non-metastatic, some forms, such as sclerodermic or terrebrant BCCs, can be highly locally invasive, leading to the severe destruction of underlying tissues and complicated surgical resections. In general, chemotherapeutic treatments of metastatic (SCC, MM), or non-metastatic but highly invasive skin cancers (BCCs), may remain to be poorly efficient when tumor cells develop resistance mechanisms due to genetic instability. Reports based on murine models have also suggested that stromal fibroblasts may contribute to the genetic instability of mammary cancer cells (Radisky et al., 2005).

Carcinoma Associated Fibroblasts (CAFs)

Combined, these observations indicate that the likelihood of curing SCC, invasive BCCs, and MM not only correlates with the size and stage of the tumors, but also with the nature of the content of the dialogue between stromal and tumor cells and its functional consequences (Gaggioli et al., 2007; Sanz-Moreno et al., 2011). Thus, significant progresses in identifying causative factors for tumor invasion and metastatic dissemination deserve further investigations in epigenetic and gene expression modifications in both tumor cells and their surrounding environment. Identification of altered pathways supporting cancer invasiveness within a permissive microenvironment would support innovative chemical targeting, taking into account the interdependance between tumor and stromal cells.

It has been known since the seventies that fibroblasts located in the vicinity of skin cancer acquire an activated phenotype similar to that observed upon healing (Kalluri and Zeisberg, 2006). For several reasons, including tissue accessibility, the possibility of fibroblasts and keratinocytes primary cell cultures, the possibility of organoid reconstruction and orthotopic grafts, skin represents an invaluable tool to identify and study the protagonists of stromal activation. In pioneer studies using in chickens infected by the Rous-Sarcoma virus, Mina Bissel and collaborators reported that skin injuries resulted in the development of invasive carcinoma in tissues that remain cancer free in the absence of such a wouding challenge. These observations suggested that the activated fibroblasts provide oncongenic signals contributing to carcinogenesis (Dolberg et al., 1985). Cat et al. (2006) reported that primary skin fibroblasts conditioned in a culture medium of SCC acquired an activated phenotype including the expression of smooth muscle actin (αSMA) and secretions of hepatocyte growth factor/scatter factor (HGF/SF), vascular endothelial growth factor (VEGF), and interleukin 6. The activation of fibroblasts is actually induced by the tumor growth factor beta 1 (TGFβ1) secreted by tumor cells. It is also known that CAFs may favor tumor growth and migration by secreting factors such as fibroblast growth factors (FGFs) and the stromal-derived factor 1α (SDF1α/CXCL12). Furthermore, activated fibroblasts provoke modifications of the extracellular matrix through the expression of matrix metallo proteinases (MMPs, notably MMP1 and MMP3 in skin) and promigratory molecules such as tenascin C (TNC). Some molecules secreted by CAFs also interact with immune cells by secreting chemotactic molecules such as monocytes chemotatic protein 1 (MCP1) and SDF1α. These molecules attract monocytes, macrophages and Natural Killer cells that, in turn, produce MMPs and growth factors impeding apoptosis and, instead, promote tumoral cell invasion (Kalluri and Zeisberg, 2006).

Recent investigations by Gaggioli and colleagues have also indicated a physical role of CAFs in pulling SCC cells within tracks to promote the collective invasion of tumor epithelial cells (Gaggioli et al., 2007). The activated fibroblast-led collective invasion of carcinoma cells required proteases (MMPs) and a force-mediated remodeling of the ECM by Rho activation of the myosin light chain (MLC). The collective invasion of SCC cells relied on the regulation of MLC by Cdc42 and MRCK (myotonin dystrophy kinase-related CDC42-binding protein kinase). Later on, it was shown that the proinflammatory cytokine IL6 of the tumor microenvironment promotes actomyosin contractility through the activation of the GP130-IL6ST receptor leading to JAK 1 kinase and Rho kinase signaling. The promotion of contractility occured in both stromal and cancer cells. Increased contractility was accompanied by the JAK1-mediated activation of the STAT3 transcription factor, suggesting a positive feedback loop to increase tumor cell invasion (Sanz-Moreno et al., 2011). Together, these observations clearly indicate the narrow interplay and co-evolution of CAF and tumor cells that result in the promotion of tumor cell invasion.

Modeling Human Cutaneous Cancer Using Primary Cells from Rare Patients Suffering from Cancer-Prone Genodermatoses

Among the numerous genodematoses that have been characterized so far, some are highly prone to skin cancers, Xeroderma pigmentosum and Nevoid Basal Cell Carcinoma (or Gorlin–Gotz) Syndromes.

Xeroderma pigmentosum is a rare recessive inherited condition due to defective nucleotide excision repair (NER). NER is a versatile DNA repair mechanism involved in the removal of mutagenic ultraviolet (UV)-induced DNA lesions such as cyclobutane pyrimidine dimers (CPD) and 6-4 photo products. In brief, NER proceeds with i-) recognition of DNA distortion provoked by UV lesions by the XPC-Rad23B complex; ii-) complex formation and stabilization of the polypeptidic repair complex around the DNA lesion (it is of interest to notice that core of this repair complex is actually made of the basal transcription factor TFIIH); iii-) TFIIH contains XPB and the XPD ATP-dependent helicases that unwind about 30 base pairs of DNA around the lesion); iv-) the ERCC1-XPF and XPG single strand endonucleases then incise the damaged DNA; v-) finally, the a DNA polymerases replicate the remaining single-strand DNA. XP is categorized into seven groups of genetic complementation (XP-A to XP-G) corresponding to a mutation in one of the XP genes (XPA to XPG) involved in NER. Patients from the XP-C group (OMIM #278720) are the most common in Europe and North Africa. XP-C patients suffer from severe photosensitivity, premature skin aging and susceptibility to develop aggressive skin cancers in the first decade of life, most notably SCC, MM and BCCs. Clinical observation indicates that the ratio of BCC to SCC, which is about 8 to 1 in the general population, is schematically reversed in XP patients. Taking this observation into account, we wondered if such clinical proneness to preferentially invasive and metastastic skin cancers could be linked to a peculiar phenotype of XP fibroblasts.

The nevoid basal cell carcinoma syndrome (NBCCS, or Gorlin's syndrome; (Gorlin, 1987) is a human genetic syndrome inherited as dominant and autosomal traits. A major clinical feature of NBCCS is a very high susceptibility to skin cancer that is limited to BCCs. Some patients also present developmental traits such as polysyndactily, bifid ribs, odontogenic keratocysts, and in about 5-10% of cases, medulloblastoma and ovarian cancers (Gorlin, 1987; Pruvost-Balland et al., 2006; Shanley et al., 1994). The disease is due to germinal mutations (about 95% nonsense) encompassing the *PATCHED1* (*PTCH1*) tumor suppressor gene. PATCHED is a putative 12-pass transmembrane receptor that binds the diffusible morphogen molecule SONIC HEDGEHOG (SHH) (Marigo et al., 1996a ; Stone et al., 1996a) (Marigo et al., 1996b; Stone et al., 1996b). Details of the pathway have been extensively described in several reviews. For not yet explained reasons, patients suffering from NBCCS present 40 to 50% BCC in non photo-exposed skin areas, suggesting that the etiology of NBCC BCCs is not necessarily directly linked to sun exposure (Goldstein et al., 1993). In addition, NBCC skin cells present normal capacities of DNA repair and UV cell survival (Brellier et al., 2008b).

On the role of Dermal Fibroblasts in Skin Cancer Susceptibility and Specificity in Xp and Nbccs

Together, clinical observations for both diseases raised questions associated with the role of the stroma:

- Why XP patients preferentially develop aggressive skin cancers (SCC, MM)?
- Why susceptibility to skin cancer in NBCCS patients is limited to BCCs?
- Why almost half NBCCS BCCs develop in non-photo exposed skin areas?

To shed light on these issues, we have developed the first alternative models of cancer-prone XP and NBCCS organotypic skin cultures *ex vivo* (OSC) (Bergoglio et al., 2007; Bernerd et al., 2005 ; Bernerd et al., 2001 ; Brellier et al., 2008a; Valin et al., 2009; Warrick et al. 2012). OSC respects two essential criteria: i-) proper 3-D organization of skin and ii-) respect of mesenchyme-epithelium interactions including the possibility to chose the genotype in each compartment (Maas-Szabowski et al., 2001).

To characterize the potential impact of *XPC* or *PATCHED* mutations in the absence of UV exposure, we first developed OSC composed of patients' epidermal cells (isolated form non-photo exposed, healthy skin biopsies), overlaying a dermal equivalent containing control fibroblasts. Under these circumstances, both types of pathologic OSCs exhibited increased numbers of Ki67-positive cells (cycling cells), an amplification of the beta 1-integrin-positive compartment, and a delayed/absent expression of keratinocyte differentiation markers (keratin 10, loricrin and filaggrin). These features were reminiscent of hyperproliferative diseases including cutaneous carcinoma. Strickingly, both XP-C and Gorlin OSC exhibited epidermal invasions in the dermal compartment (Bernerd et al., 2005; Bernerd et al., 2001). NBCCS epidermal keratinocytes showed an over expression of the GLI2 transcription factor and accumulation of nuclear Cyclin D1 as in autologous patients' BCCs, suggesting a constitutive activation of the SHH pathway in primary NBCCS cells (Brellier et al., 2008a).

Conversely, OSC composed of a pathologic dermal equivalent, containing either XP-C or NBCCS primary fibroblasts underlying a control epidermis, also exhibited features of epidermal invasion and dermo epidermal clefts (characteristic of BCCs), respectively. We then produced fully pathological OSC, in which both skin compartments contained patient's keratinocytes and fibroblasts, respectively. Under these circumstances, differentiation defects worsened.

In XP fibroblasts, we studied the expression of candidate genes whose products are responsible for extracellular matrix (ECM) modifications upon cancer, chronological aging, and photo aging (Shin et al., 2005). Nine out of nine independent XP-C primary fibroblast strains revealed an overexpression of MMP1 and, to a lesser extent, of MMP3 (collagenase I, stromelysin, respectively). We demonstrated that the MMP1 overexpression resulted from the abnormal activation of the AP1 transcription pathway, with a notable FOSB overexpression (Frechet et al., 2008). This resulted from an accumulation of reactive oxygen species (ROS) in XP-C primary fibroblasts. Elimination of ROS accumulation using Epigallo Catechin Gallate (EGCG) led to MMP1 normalization. Levels of ROS remained stable over serial propagation of cells in the presence of EGCG and no toxicity could be observed under our experimental conditions. Insofar, as the management of XP patients remains limited to photo protection and surgery, these results are of potential interest for using green tea polyphenol derivatives in the pharmacological treatment of XP patients.

Role of the Microenvironment in the Physiopathology of the Gorlin Syndrome

For several decades, experimental evidences have suggested that BCCs were highly dependent on their microenvironment. In the sixties, it had been shown that the take and survival of sporadic BCCs grafted on human patients from the general population were highly dependent upon the presence of dermal cells (Lyles et al., 1960). The take of BCC was even better in patients suffering from the Gorlin syndrome. Although such experiments are ethically unconceivable today, they strongly supported the idea of a stromal dependence of BCCs cells (Blackwell and McComb, 1969).

Other evidences support this idea. *In vivo*, BCCs exceptionally give rise to distal metastases. Survival of BCC keratinocytes could thus depend on their local dermal environment. This could be due to the reduced plasticity of keratinocytes from which BCCs develop as compared to SCC-founding keratinocytes. In accordance with this hypothesis, it is worth noting that human BCC cells can hardly be cultured *in vitro*, whereas SCC cells can be grown easily.

BCCs Primary Fiboblasts Express a Phenotype Similar to That of CAF from Sporadic BCCs

To further characterize the phenotype of NBCCs primary skin fibroblasts we performed analyses of the whole genome expression in six strains isolated from non-photo-exposed skin (buttock) bearing unrelated mutations (3 missense; 3 nonsense). Three independant strains of primary fibroblasts isolated from individuals from the general population were used as controls. In order to respect the 3D architecture of skin, fibroblasts were embeded into collagen lattices before gene expression analyses. Under these circumstances, a signature of 18 genes was identified and their expression profile was highly similar to that described in CAFs from sporadic BCCs (Valin et al., 2009).

Table 1. The expression of the genes involved in cell proliferation, differentiation, migration, extracellular cell matrix remodelling, chemiotactism, and angiogenesis

Gene	Fold change
MMP1	8.4 (p<0.025)
MMP3	26.8 (p<0.025)
COL3A1	0.7 (p<0.05)
COL7A1	0.6 (p<0.05)
COL11A1	11.5
LAMA2	0.4 (p<0.025)
TNC	1.2
CXCL12	2.4 (p<0.025)
MGP	5.2 (p<0.025)
ANGPTL2	3 (p<0.025)
ANGPTL4	9.8 (p<0.025)
FGF7	2.3 (p<0.05)
GREM1	1.7
SFRP2	3.5 (p<0.025)
DKK3	0.3 (p<0.025)
WNT5A	0.5 (p<0.05)
WISP2	2.7 (p<0.05)
ID2	2.6 (p<0.025)

These 18 genes' products are involved in cell proliferation, differentiation, migration, extracellular cell matrix remodelling, and chemiotactism as well as angiogenesis (see table 1). Strikingly, MMP1 and MMP3 mRNAs were increased by 8.4 and 26.8 (p< 10-5), respectively. Conversely, some components of the ECM, such as collagen 7A1, collagen 3A1, and Laminin2, were found significantly decreased (0.6, 0.6 and 0.4, respectively). Our results also indicated that the genes already known to be over expressed in CAF from sporadic BCCs (SDF1α, TNC, and Gremlin 1) and known to promote tumor cell proliferation and invasion were also over expressed in NBCCS fibroblasts. Interestingly, SDF1α is degraded by MMP1 and MMP3. SDF1α is over expressed in breast cancer, but its chemical inhibition results in decreased numbers of metastases. Thus, the degradation of SDF1α by MMP1 could contribute to the poor metastatic potential of BCCS. Finally, our finding that shows a significant ovexpression of ANGPTL4, an antiangiogenic agent, may also limit the growth and invasion of BCCs as compared to SCCs appearing much more vascularized. This remains a hypothesis insofar as expression levels of ANGPTL4 have not been measured in sporadic BCCs. These observations are summerized in Figure1.

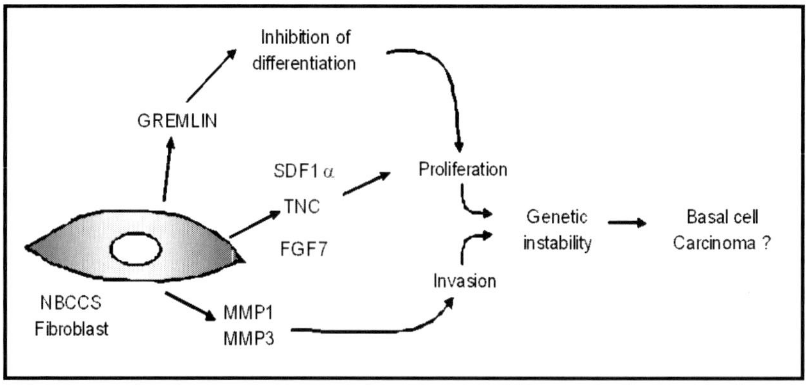

Figure 1. Schematic represention of putative effects of NBCCS fibroblasts on epidermal keratinocytes.

Conclusion

Based on our model systems and findings, innovative pharmacological approaches should emerge that will not only benefit to both XP and NBCCs patients, but also to patients from the general population suffering from invasive cancers. In spite of these interesting perspectives, we are still missing the molecular mechanisms connecting the absence of XPC and its impact on genome expression. Notably, the overexpression of MMP1 and other proinvasive factors are currently under study. A molecular link is also missing in between PATCHED 1 +/- NBCCS cells and the alterations of the expression of their genome. Further studies should fill this gap.

Acknowledgments

This work was supported by the Fondation de l'Avenir (ET9-551), the AFM (1315; 15521), the Association pour la Recherche sur le Cancer (SFI201212055859). We are indebted to Drs Florence Brellier, Alexandre Valin, Stéphanie Barnay, Mathilde Frechet, Emilie Warrick, Sabine Scarzello, Odile Chevallier-Lagente, Pr. Alain Sarasin, and Pr. Marie-Françoise Avril for their expert and enthousiastic contribution to this work.

References

Bergoglio, V., Larcher, F., Chevallier-Lagente, O., Bernheim, A., Danos, O., Sarasin, A., Rio, M. D., and Magnaldo, T. (2007). Safe selection of genetically manipulated human primary keratinocytes with very high growth potential using CD24. *Mol. Ther. 15*, 2186-2193.

Bernerd, F., Asselineau, D., Frechet, M., Sarasin, A., and Magnaldo, T. (2005). Reconstruction of DNA repair-deficient xeroderma pigmentosum skin in vitro: a model to study hypersensitivity to UV light. *Photochem. Photobiol. 81*, 19-24.

Bernerd, F., Asselineau, D., Vioux, C., Chevallier-Lagente, O., Bouadjar, B., Sarasin, A., and Magnaldo, T. (2001). Clues to epidermal cancer proneness revealed by reconstruction of DNA repair-deficient xeroderma pigmentosum skin in vitro. *Proc. Natl. Acad. Sci. USA 98*, 7817-7822.

Blackwell, J. B., and McComb, H. (1969). The long-term results of autologous transplantation of basal cell carcinoma in skin grafts. *Cancer 23*, 101-108.

Brellier, F., Bergoglio, V., Valin, A., Barnay, S., Chevallier-Lagente, O., Vielh, P., Spatz, A., Gorry, P., Avril, M. F., and Magnaldo, T. (2008a). Heterozygous mutations in the tumor suppressor gene PATCHED provoke basal cell carcinoma-like features in human organotypic skin cultures. *Oncogene 27*, 6601-6606.

Brellier, F., Valin, A., Chevallier-Lagente, O., Gorry, P., Avril, M. F., and Magnaldo, T. (2008b). Ultraviolet responses of Gorlin syndrome primary skin cells. *Br. J. Dermatol. 159*, 445-452.

Cat, B., Stuhlman, D., Steinbrenner, H., Alili, L., Holtkötter, O., Sies, H., and Brenneisen, P. (2006). Enhancement of tumor invasion depends on transdifferentiation of fibroblasts mediated by rapid oxygen species. *J. Cell Sci. 119*, 2727-2738.

Dolberg, D. S., Hollingsworth, R., Hertle, M., and Bissell, M. J. (1985). Wounding and its role in RSV-mediated tumor formation. *Science 230*, 676-678.

Frechet, M., Warrick, E., Vioux, C., Chevallier, O., Spatz, A., Benhamou, S., Sarasin, A., Bernerd, F., and Magnaldo, T. (2008). Overexpression of matrix metalloproteinase 1 in dermal fibroblasts from DNA repair-deficient/cancer-prone xeroderma pigmentosum group C patients. *Oncogene 27*, 5223-5232.

Gaggioli, C., Hooper, S., Hidalgo-Carcedo, C., Grosse, R., Marshall, J. F., Harrington, K., and Sahai, E. (2007). Fibroblast-led collective invasion of carcinoma cells with differing roles for RhoGTPases in leading and following cells. *Nat. Cell Biol. 9*, 1392-1400.

Goldstein, A. M., Bale, S. J., Peck, G. L., and DiGiovanna, J. J. (1993). Sun exposure and basal cell carcinomas in the nevoid basal cell carcinoma syndrome. *J. Am. Acad. Dermatol. 29*, 34-41.

Gorlin, R. J. (1987). Nevoid basal-cell carcinoma syndrome. *Medicine* (Baltimore) *66*, 98-113.

Kalluri, R., and Zeisberg, M. (2006). Fibroblasts in cancer. *Nat. Rev. Cancer 6*, 392-401.

Lyles, T. W., Freeman, R. G., and Knox, J. M. (1960). Transplantation of basel cell epitheliomas. *J. Invest Dermatol. 34*, 353.

Maas-Szabowski, N., Szabowski, A., Stark, H. J., Andrecht, S., Kolbus, A., Schorpp-Kistner, M., Angel, P., and Fusenig, N. E. (2001). Organotypic cocultures with genetically modified mouse fibroblasts as a tool to dissect molecular mechanisms regulating keratinocyte growth and differentiation. *J. Invest Dermatol. 116*, 816-820.

Marigo, V., Davey, R. A., Zuo, Y., Cunningham, J. M., and Tabin, C. J. (1996a). Biochemical evidence that patched is the Hedgehog receptor. *Nature 384*, 176-179.

Marigo, V., Davey, R. A., Zuo, Y., Cunningham, J. M., and Tabin, C. J. (1996b). Biochemical evidence that patched is the Hedgehog receptor. *Nature 384*, 176-179.

Polyak, K., Haviv, I., and Campbell, I. G. (2009). Co-evolution of tumor cells and their microenvironment. *Trends Genet 25*, 30-38.

Pruvost-Balland, C., Gorry, P., Boutet, N., Magnaldo, T., Mamelle, G., Margulis, A., Kolb, F., Duvillard, P., Spatz, A., Brugieres, L., *et al.* (2006). [Clinical and genetic study in 22 patients with basal cell nevus syndrome]. *Ann. Dermatol. Venereol. 133*, 117-123.

Radisky, D. C., Levy, D. D., Littlepage, L. E., Liu, H., Nelson, C. M., Fata, J. E., Leake, D., Godden, E. L., Albertson, D. G., Nieto, M. A., *et al.* (2005). Rac1b and reactive oxygen species mediate MMP-3-induced EMT and genomic instability. *Nature 436*, 123-127.

Sanz-Moreno, V., Gaggioli, C., Yeo, M., Albrengues, J., Wallberg, F., Viros, A., Hooper, S., Mitter, R., Feral, C. C., Cook, M., *et al.* (2011). ROCK and JAK1 Signaling Cooperate to Control Actomyosin Contractility in Tumor Cells and Stroma. *Cancer Cell 20*, 229-245.

Shanley, S., Ratcliffe, J., Hockey, A., Haan, E., Oley, C., Ravine, D., Martin, N., Wicking, C., and Chenevix-Trench, G. (1994). Nevoid basal cell carcinoma syndrome: review of 118 affected individuals. *Am. J. Med. Genet. 50*, 282-290.

Shin, M. H., Rhie, G. E., Kim, Y. K., Park, C. H., Cho, K. H., Kim, K. H., Eun, H. C., and Chung, J. H. (2005). H2O2 accumulation by catalase reduction changes MAP kinase signaling in aged human skin in vivo. *J. Invest Dermatol. 125*, 221-229.

Stone, D. M., Hynes, M., Armanini, M., Swanson, T. A., Gu, Q., Johnson, R. L., Scott, M. P., Pennica, D., Goddard, A., Phillips, H., *et al.* (1996a). The tumour-suppressor gene patched encodes a candidate receptor for Sonic hedgehog. *Nature 384*, 129-134.

Stone, D. M., Hynes, M., Armanini, M., Swanson, T. A., Gu, Q., Johnson, R. L., Scott, M. P., Pennica, D., Goddard, A., Phillips, H., *et al.* (1996b). The tumour-suppressor gene patched encodes a candidate receptor for Sonic hedgehog. *Nature 384*, 129-134.

Valin, A., Barnay-Verdier, S., Robert, T., Ripoche, H., Brellier, F., Chevallier-Lagente, O., Avril, M. F., and Magnaldo, T. (2009). PTCH1 +/- dermal fibroblasts isolated from healthy skin of Gorlin syndrome patients exhibit features of carcinoma associated fibroblasts. *PLoS One 4*, e4818.

Warrick, E., Garcia, M., Chagnoleau, C., Chevallier, O., Bergoglio, V., Sartori, D., Mavilio, F., Angulo, J. F., Avril, M. F., Sarasin, A., *et al.* (2012). Preclinical Corrective Gene Transfer in Xeroderma Pigmentosum Human Skin Stem Cells. *Mol. Ther. 20:* 798-807.

In: Dermal Fibroblasts
Editor: Xiaowen Bai
ISBN: 978-1-62618-799-3

Chapter 7

Dermal Fibroblasts in Pathomechanism of Pachydermoperiostosis

Jun-ichi Sakabe,[1] Yoshiki Tokura[1] and Kenji Kabashima[2]
[1] Department of Dermatology,
Hamamatsu University School of Medicine, Hamamatsu, Japan
[2] Department of Dermatology,
Kyoto University Graduate School of Medicine, Kyoto, Japan

Abstract

The dermis is a dynamic, supportive connective tissue composed of an extracellular matrix (collagen, elastic fibers, hyaluronic acids, etc.) and ground substance that is mainly formed by dermal fibroblasts. Dermal fibroblasts interact with each other and further with keratinocytes, dendritic cell, monocytes and/or macrophages, neutrophils and lymphocytes, playing a crucial role in regulating skin physiology, such as wound repair and anti-aging.

On the other hand, an aberrant fibroblast phenotype is known to contribute to the pathogenesis of fibrotic disorders, such as systemic sclerosis, localized scleroderma, nephrogenic systemic fibrosis, and keloid scars.

It has recently been revealed that the etiology of a rare genetic disease, pachydermoperiostosis (PDP) or primary hypertrophic osteoarthropathy (PHO), is related to abnormalities in dermal fibroblasts. PDP or PHO is characterized by their unique phenotypes of the skin and bone, such as thick skin and periostitis of long bones. PDP/PHO is attributed to the mutations in the *15-hydroxyprostaglandin dehydrogenase* (*15-PGDH*) gene or the *solute carrier organic anion transporter family, member 2A1* (*SLCO2A1*) gene, which lead to elevated serum prostaglandin E_2. The aim of this chapter is to review the roles of dermal fibroblasts in the pathogenesis of pachydermia in association with Wnt/β-catenin signal pathway and prostaglandin E_2. Enhanced Wnt/β-catenin signaling induced by prostaglandin E_2 seems to contribute to the development of pachydermia by enhanced dermal fibroblast functions.

Keywords: Wnt/β-catenin signal pathway, pachydermoperiostosis (PDP), primary hypertrophic osteoarthropathy (PHO), prostaglandin E_2 (PGE_2), *15-PGDH*, *SLCO2A1*

Introduction

Pachydermoperiostosis (PDP), or primary hypertrophic osteoarthropathy (PHO), is a rare disease (Rimoin, 1965; Shawarby and Ibrahim, 1962; Vogl and Goldfischer, 1962) diagnosed by the presence of a triad of pachydermia (skin thickening), digital clubbing, and periostosis of long bones. Typically, insidious development of thickening of the fingers and toes, clubbing of the terminal phalanges, enlargement of the hands and feet, hyperhidrosis, increased sebaceous secretion, and velvet coloration of the skin occur mostly in men during adolescence (Jajic et al., 2001).

Radiographic signs of bilateral and symmetrical periostosis are frequently observed as a marked irregular periosteal ossification of the tibias and fibulas (Rimoin, 1965). Touraine et al. recognized PDP with three clinical presentations or forms: a 'complete form' presenting the full-blown phenotype, an 'incomplete form' characterized by the phenotype without pachydermia, and a 'fruste form' with pachydermia and minimal or absent skeletal changes (Touraine et al., 1935).

Recently, PDP/PHO was mapped to chromosome 4q33-q34, and gene mutations in *HPGD,* encoding 15-hydroxyprostaglandin dehydrogenase, the main enzyme of prostaglandin (PG) degradation, were identified (Uppal et al., 2008). In addition, mutations in the *solute carrier organic anion transporter family, member 2A1* (*SLCO2A1*) gene on chromosome 3q22.1-q22.2, encoding PG transporter were also detected in PDP/PHO (Sasaki et al., 2012; Zhang et al., 2012; Seifert et al., 2012; Diggle et al., 2012; Busch et al., 2012). Since these mutations lead to the elevation of serum PGE_2, we hypothesized that PGE_2 may be involved in the pathogenesis of pachydermia in PDP. Since the major manifestations of complete PDP occur in both skin and bone, the etiology could be related to the dysregulation of modulators for mesenchymal cells, such as bone morphogenetic proteins (BMP), transforming growth factor (TGF)-β and factors involved in wingless (Wnt) pathways (Botchkarev and Sharov, 2004; Castori et al., 2005; Kornak and Mundlos, 2003). The Wnt signaling consists of canonical and non-canonical pathways. The canonical pathway involves cytosolic β-catenin stabilization, nuclear translocation and gene regulation, and the non-canonical pathways activate rho, rac, JNK and protein kinase C (Kawano and Kypta, 2003; Zorn, 2001). These signaling pathways are mediated by Wnt protein, which binds to a frizzled Wnt receptor. Wnt signaling is modulated by several different families of secreted downregulators.

Among them, Dickkopf (DKK) is a family of cysteine-rich proteins comprising at least four different forms (DKK1, DKK2, DKK3, and DKK4), which are coordinately expressed in mesodermal lineages. The best studied of these is DKK1, which blocks the canonical Wnt signaling by inducing endocytosis of lipoprotein receptor-related protein 5/6 (LRP5/6) complex (Mao et al., 2002) without affecting the frizzled Wnt receptor (Mao et al., 2001). DKK1 induces the formation of ectopic heads in *Xenopus laevis* in the presence of BMP inhibitors (Glinka et al., 1998) and modulates apoptosis during vertebrate limb development (Grotewold and Ruther, 2002). High mRNA levels of *DKK1* in human dermal fibroblasts of the palms and soles inhibit the function and proliferation of melanocytes via the suppression of β -catenin and microphthalmia-associated transcription factor (MITF) (Yamaguchi et al., 2004; Yamaguchi et al., 2007). In parallel, DKK1 transgenic mice under the control of keratin 14 have no pigmentation on the trunk because of the absence of melanocytes in the inner-follicular epidermis as well as the lack of hair follicle development (Andl et al., 2002). These findings suggest that DKK1 is deeply involved in the formation and differentiation of the skin.

Here we investigated two complete cases of PDP using dermal fibroblasts to address the pathogenesis. Profiles of mRNA expression in cultured dermal fibroblasts were examined by DNA microarray analysis and real-time RT-PCR. DKK1 and β-catenin protein expressions were evaluated by immunohistochemistry in the skin.

To evaluate the in vivo roles of DKK1 in mice, *DKK1* siRNA was injected to the ears. PDP fibroblasts proliferated more than control fibroblasts. mRNA expression of a Wnt signaling antagonist, *DKK1*, was much lower in PDP fibroblasts than in normal ones. Consistently, decreased expression of DKK1 in fibroblasts and enhanced expression of β-catenin were noted in PDP patients.

Moreover, recombinant human DKK1 protein decreased the proliferation of dermal fibroblasts. Furthermore, PGE_2 decreased DKK1 mRNA expression and increased β-catenin mRNA and protein levels in normal human dermal fibroblasts.

Therefore, PGE_2 is thought to be an important factor in the pathogenesis of PDP/PHO via activation of Wnt/β-catenin signal. In accord with the above human studies, intradermal injections of *DKK1* siRNA into mouse ears increased ear thickness as seen in PDP. These results suggest that enhanced Wnt signaling contributes to the development of pachydermia by enhanced dermal fibroblast functions.

Methods

Patients

Case 1. A 50-year-old male was referred to our clinic as reported previously (Matucci-Cerinic et al., 1992). The skin on his head and face was thick and oily with a dark velvet color. Naso-labial folds and transverse furrowing of the forehead were prominent (Figure 1. A).

The hands were enlarged with marked clubbing of the second and fifth digits compared to those of an age- and sex-matched healthy donor (Figure 1. B). These symptoms developed when he was 18 years old. X-ray examination of the long bones showed major periostosis with cortical thickening and widening of the shafts (data not shown).

Histology of the skin showed thickened dermis, and sebaceous and sweat gland enlargement compared to that of a healthy control (Figure 1. C). Elastica van Gieson staining showed thick and interwoven collagen bundles in some areas of the dermis and also thick and partially fragmented elastic fibers in PDP (Figure 1. D). The intensity of mucinous ground substance observed by alcian blue staining was comparable between a healthy control and a PDP patient (Figure 1. E). Neither hepatosplenomegaly nor internal malignancy was found on physical examination or CT scans. Biochemical tests showed normal levels of thyroid-stimulating hormone and growth hormone, which likely rules out thyroid acropathy and acromegaly.

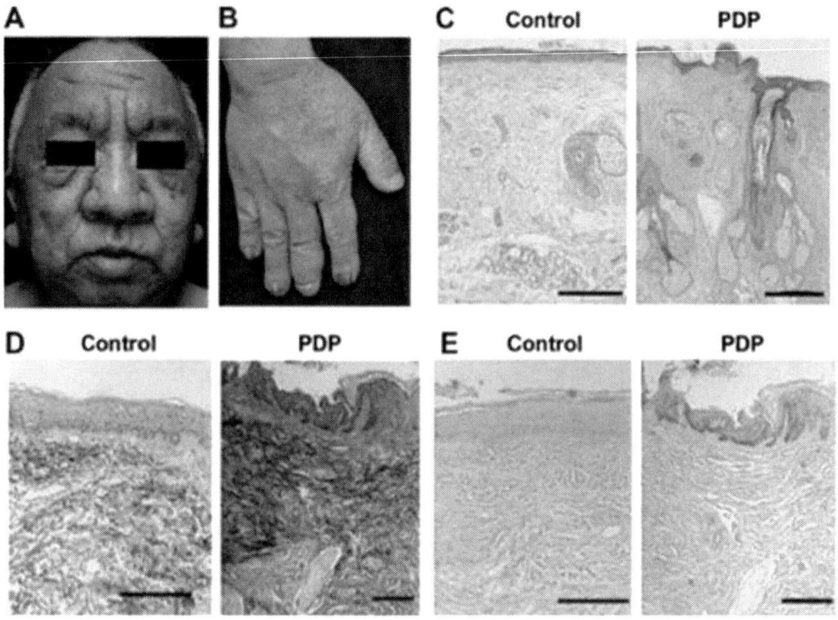

Figure 1. Clinical findings and histology of a PDP patient. (A, B) Clinical appearances of the face (A) and hands (B) in PDP (case 1) are shown. (C) Histology of the face in PDP shows thickened dermis, and hyperplasia of the sebaceous and sweat glands by HandE staining compared to a healthy donor (control). Scale bars, 300 μm. (D) The sample of PDP stained with Elastica van Gieson shows thick collagen and elastic fibers in the dermis compared to a healthy donor. (E) The intensity of mucinous ground substance observed by alcian blue staining is comparable between a healthy donor and a PDP patient. Scale bars, 100 μm (D, E) (adapted from ref. (Kabashima et al., 2010)).

Family history was noncontributory. Based on these clinical manifestations and histological findings, the patient was diagnosed as the complete form of PDP. In this case, an *HPGD* mutation was not found, and an *SLCO2A1* gene mutation was not examined thus far.

Case 2. The patient was a 38-year-old male with clinical findings similar to case 1, including pachydermia, digital clubbing, and periostosis. He had no signs or symptoms of hepatosplenomegaly, pulmonary diseases, tumoral syndrome, thyroid acropathy, or acromegaly (data not shown) as reported previously (Masuda et al., 2000).

Cell Preparation, Culture, and Reagents

Skin biopsies of the right temple (case 1) and scalp (case 2) were performed for histology and primary culture of fibroblasts. Control donors were matched for age, sex, and biopsy site, and the samples were processed in parallel. Institutional approval and informed consent were obtained from all subjects. The biopsy samples were immersed in Dulbecco's Modified Eagle Medium (Sigma, St. Louis, MO) containing 10% heat-inactivated fetal calf serum (Invitrogen, Carlsbad, CA), 5×10^{-5} M 2-mercaptoethanol, 2 mM L-glutamine, 25 mM HEPES (Cellgro, Herndon, VA), 1 mM nonessential amino acids, 1 mM sodium pyruvate,

100 units/ml penicillin, and 100 μg/ml streptomycin, with 5% CO_2 at 37°C. The fibroblasts were allowed to adhere to the surface of 100-mm plastic tissue culture dishes (Nunc, Roskilde, Denmark). To evaluate the number of fibroblasts, 2×10^5 3rd-passage fibroblasts were seeded in 1 ml of medium in 24-well dishes and resuspended with trypsin/EDTA one week later. The numbers of fibroblasts were evaluated 7 and 14 days after seeding by flow cytometry using FACSCanto (BD Biosciences, San Diego, CA) with standard beads Flow Count (Beckman Coulter, Fullerton, CA). The actin bundle formation of cultured fibroblasts from a healthy individual (control) and an individual with PDP (PDP) were examined by staining with alexa 488-labeled phalloidin antibody (Invitrogen) 5 days after the 4th passage. For treatment with DKK1, fibroblasts were harvested 5 days after a comparable number of passages and cultured again at 1×10^6 cells in one ml of medium with or without recombinant human DKK1 (RandD Systems Inc., Minneapolis, MN) for another 2 days. For treatment with PGE_2, fibroblasts were harvested 5 days after a comparable number of passages and cultured again at 1×10^5 cells in 2 ml of medium with or without PGE_2 (Sigma) in the presence of indomethacin (10 μM; Cayman Chemical Co., Ann Arbor, MI) for another 4 days.

Flow Cytometry and Histology

Flow cytometric analysis was performed with doublet discrimination on the FACSCanto (Kabashima et al., 2005) and FlowJo software (TreeStar, San Carlos, CA) (Kabashima et al., 2006). Human fibroblasts were treated with cytofix/cytoperm buffer according to the manufacturer's protocol (BD Biosciences). For cell cycle analysis, fibroblasts were incubated with 7-amino actinomycin D (7-AAD) (BD Biosciences) for 20 min at 4°C. After staining with 7-AAD, the DNA contents were analyzed by flow cytometry. For β-catenin staining, fibroblasts were stained with phycoerythrin-labeled β-catenin antibody (H-102, Santa Cruz Biotechnology Inc., Santa Cruz, CA), and mean fluorescence intensity was evaluated by flow cytometry.

For histology, the biopsy samples and the ears of mice were fixed in 10% formaldehyde. Sections of 5-μm thickness were prepared and stained with hematoxylin and eosin, Elastica van Gieson, or alcian blue. Immunohistochemical staining on paraffin-embedded sections was performed using a Vectastain ABC kit (Vector Laboratories, Burlingame, CA) (Kabashima et al., 2005). Antibodies used were rabbit anti-human polyclonal DKK1 (ab61034, Abcam, Cambridge, UK), mouse monoclonal anti-human β-catenin IgG1 (610153, BD Biosciences, San Diego, CA), and rabbit anti-human polyclonal proliferating cellular nuclear antigens (PCNA) antibodies (SC-7907, Santa Cruz Biotechnology Inc., Santa Cruz, CA). The control antibodies used were rabbit non-immune serum or mouse IgG1 (X0931, Dako, Glostrup, Denmark). The immunoreactivity was visualized by Fast Red or diaminobenzidine (Sigma), and the sections were then counterstained with hematoxylin.

Western Blot Analysis

For western blotting studies, fibroblasts were isolated from a healthy donor. Cytoplasm- and nuclear- proteins were extracted by NucBuster™ Protein Extraction Kit (Novagen,

Darmstadt, Germany). Twenty μg protein samples were electrophoresed by 8% SDS-PAGE and electroblotted onto PVDF membranes for 2 h at 180 mA. After blocking with 5% skim milk solution, the membranes were incubated with rabbit anti-human β-catenin (SC-7199; 1:1000, Santa Cruz Biotechnology Inc.) polyclonal antibodies or rabbit anti-human GAPDH (SC-25778; 1:1000, Santa Cruz Biotechnology Inc.) antibody and detected with horseradish peroxidase-conjugated goat anti-rabbit IgG (Bio-Rad, Hercules, CA). Immunoblots were visualized using the ECL Plus Western Blotting Detection Reagents (GE Healthcare, Buckinghamshire, UK) according to the manufacturer's protocol. Bands were quantified by densitometry with the help of a CS Analyzer ver. 2.0 (ATTO, Tokyo, Japan).

Quantitative RT-PCR and Microarray Procedures

Total RNA was extracted from three-passage fibroblasts (case 1 and the control) cultured for 2 days with the RNeasy Mini Kit (QIAGEN, Valencia, CA). cDNA was reverse transcribed from total RNA samples using the TaqMan RT reagents (Applied Biosystems, Foster City, CA). Human *DKK1* (Assay ID: Hs00183740) mRNA expression was quantified using TaqMan Gene Expression Assay (Applied Biosystems) with the ABI PRISM 7700 sequence detection system (Applied Biosystems). As an endogenous reference for these RT-PCR quantification studies, human *GAPDH* control reagents (Assay ID: Hs99999905) (Applied Biosystems) were used. The relative expression was calculated using the delta-delta Ct method (Livak and Schmittgen, 2001).

For DNA microarray analysis, total RNAs were extracted from fibroblasts with the RNeasy Mini Kit (QIAGEN). For transcriptomic profiling, we used an oligonucleotide-based DNA microarray, AceGene (HumanOligoChip30K, DNA Chip Research, Yokohama, Japan). Images were analyzed with DNASIS Array (Hitachi Software Engineering, Tokyo, Japan), according to the manufacturer's instructions. Mean and standard deviation (SD) of background levels were calculated, and genes with intensities less than mean plus 2SD of background levels were excluded from further analysis. The Cy5/Cy3 ratios of all spots on the DNA microarray were normalized by the method of global normalization.

Application of Mouse *DKK1* siRNA

Mouse *DKK1* siRNA (5'-gaa cca cac uga cuu caa aTT - 3') was purchased from Nippon EGT (Toyama, Japan). siRNA duplexes were generated by mixing sense and antisense single-strand RNA oligomers equally in an annealing buffer (NIPPON EGT) (Maekawa et al., 2005). Negative control siRNA (AM4611) was purchased from Ambion (Austin, TX). To impregnate mouse *DKK1* siRNA into cationized gelatin microspheres (Kushibiki et al., 2004), 10 μl of PBS solution (pH 7.4) containing 10 μg of mouse *DKK1* siRNA was dropped onto 1 mg of the freeze-dried cationized gelatin microspheres, kept overnight at 4°C, and added to 190 μl of PBS. Ten μl of this siRNA solution was injected intradermally into the center of the ears of 8 week-old C57BL/6j female mice (obtained from SLC, Shizuoka, Japan) using a 30-gauge needle 4 times every 7 days. The same amount of cationized gelatin-conjugated non-sense siRNA was applied as a negative control. The ear thickness was measured before each injection and one week after the last injection using dial-thickness gauge (PG-01, TECLOCK,

Okaya, Japan). The injected area was sampled for histology and RT-PCR analysis using 6 mm punch biopsy. Mice were maintained on a 12-hour light/dark cycle under specific pathogen-free conditions. Protocols were approved by the Institutional Animal Care and Use Committee of the University of Occupational and Environmental Health.

Statistical Analysis

Data were analyzed using an unpaired two-tailed *t*-test. A *P*-value of less than 0.05 was considered to be significant.

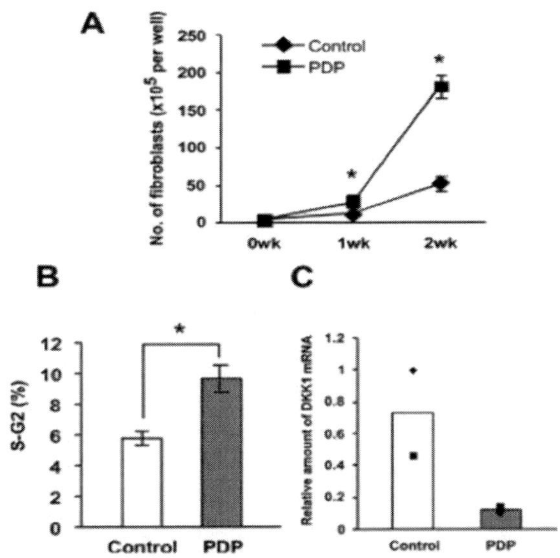

Figure 2. Characteristics of dermal fibroblasts and histology of the skin in PDP. (A) Fibroblasts from a healthy individual (control) and an individual with PDP (case 1) (PDP) were incubated and the numbers of fibroblasts examined. (B) The percentages of fibroblasts in S-G2 phase are shown. (C) The levels of *DKK1* mRNA in fibroblasts from two controls and two PDPs were normalized against GAPDH, and the level of one of the control *DKK1* mRNAs is regarded as one. Filled symbols indicate two independent individuals and columns represent the average. The student's *t* test was performed ($P < 0.05$) (A, B) (adapted from ref. (Kabashima et al., 2010)).

Result

Increased S-G2 Phase in Fibroblasts of PDP

Case 1 had a typical complete form of PDP (Figures 1. A and B) characterized by the triad of pachydermia, digital clubbing and periostosis (Rimoin, 1965; Shawarby and Ibrahim, 1962; Vogl and Goldfischer, 1962). The histology of the skin showed thickened dermis with dense and packed collagen and elastic fibers (Figures 1. C-E), suggesting that the function of fibroblasts was enhanced in PDP. To test the proliferative activity of fibroblasts, we cultured primary fibroblasts from case 1 and a matched control, and monitored their number. As reported previously (Matucci-Cerinic et al., 1992), the number of PDP fibroblasts was significantly higher than that of control fibroblasts (Figure 2. A). Similar results were obtained in another typical patient with PDP, case 2 (data not shown). To clarify whether it was due to enhanced cell survival or proliferation, we stained the nuclear contents of fibroblasts with 7-AAD for cell cycle analysis. The ratio of PDP fibroblasts in the cell cycle (S-G2 phase) was higher than that of control fibroblasts (Figure 2. B), suggesting that the proliferation of fibroblasts was enhanced in PDP.

Decreased DKK1 Expression in PDP Fibroblasts and Skin

The above results together with the clinical phenotypes involving the skin and bone suggested the possibility that the pathogenesis of PDP is related to dysregulation of BMP, Wnt, and/or TGF-β pathways in mesenchymal cells. To efficiently compare the expression profiles of these genes between PDP fibroblasts (case 1) and matched controls, DNA microarray analysis was performed and the complete array data was deposited in a MIAME-compliant microarray database (GSE17947). Among all genes analyzed, 2573 genes were elevated and 2346 genes were decreased more than 2 folds in PDP patients compared to a healthy control. The analysis revealed that the mRNA levels of *BMP* and *TGF-β* families were comparable between these two groups (Kabashima et al., 2010). On the other hand, *WNT1*, *WNT10B*, and *DKK1* mRNAs were decreased in the patient's fibroblasts. In particular, *DKK1* mRNA was markedly decreased. Other molecules, such as levels of *LRP5/6*, *Kremen1*, and *Kremen2* mRNA were similar between these two groups. Moreover, the mRNA levels for collagen families, such as *COL4A5, COL9A2, COL10A1, COL11A1, COL12A1, COL14A1*, and *COL17A1*, were elevated, but those for *fibronectin* and *elastin* (*ELN*) families were not (Kabashima et al., 2010). These data suggest that the PDP fibroblasts showed enhanced production of several types of collagens in addition to cell proliferation, which might explain the pathogenesis of pachydermia in PDP.

We initially confirmed the decreased *DKK1* expression using quantitative RT-PCR. Fibroblasts were primarily cultured from two PDP patients (cases 1 and 2) and two matched healthy controls. *DKK1* mRNA levels in PDP fibroblasts were consistently lower than those in the control fibroblasts (Figure 2. C). We then performed immunohistochemical analysis to evaluate the expression of DKK1 protein in the PDP skin (case 1) and the control. In the normal skin, DKK1 was detected diffusely in the dermis and notably in the cytoplasm of fibroblasts. The intensity of this expression pattern was substantially decreased in the PDP patient (case 1) (Kabashima et al., 2010). This finding was confirmed with the other PDP patient (case 2) and another matched control (data not shown). We displayed that the controls incorporating non-immune serum or mouse IgG1 showed no specific reactivity (Kabashima et al., 2010).

The decreased expression of DKK1 in PDP suggests that Wnt signaling is enhanced in PDP. Immunohistochemical analysis revealed enhanced β-catenin expression in the PDP skin (case 1), especially around the sebaceous glands, the hair follicles, and the epidermis, and mildly in the dermis compared to the control (Kabashima et al., 2010), supporting the augmented expression of Wnt signaling.

Suppression of Fibroblast Proliferation by DKK1

The above results indicated that Wnt signaling is enhanced in PDP through decreased DKK1 expression. However, it was still unknown whether DKK1 directly modulates the function of dermal fibroblasts. To solve this issue, we cultured dermal fibroblasts from a healthy control and the patient with PDP (case 1) in the presence or absence of human recombinant DKK1, and quantitated the DNA contents of fibroblasts by cell cycle analysis with 7-AAD. The ratio of fibroblasts in the cell cycle (S-G2 phase) was higher in the PDP patient than in the control (Figure 3. A, B). In addition, the ratio of fibroblasts with the cell cycle (S-G2 phase) was decreased by treatment with recombinant DKK1 protein in a dose-dependent manner (Figure 3. B), implicating the direct involvement of DKK1 in fibroblast proliferation.

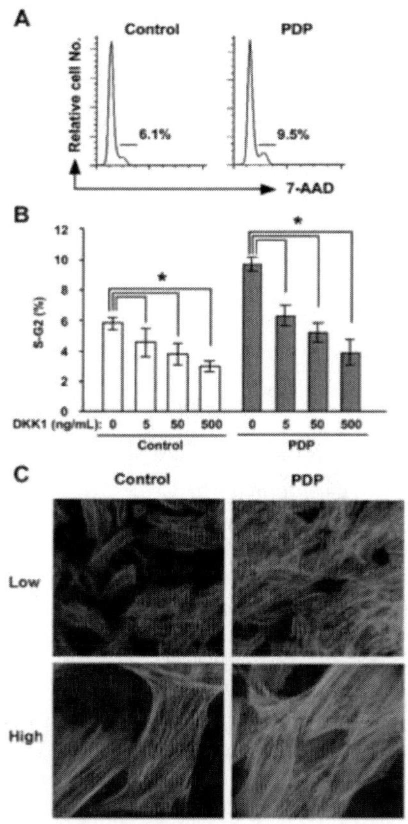

Figure 3. The effect of DKK1 on fibroblast proliferation and actin bundle formation of fibroblasts. (A, B) The fibroblasts from a healthy individual (control) and an individual with PDP (PDP) were incubated with or without recombinant human DKK1 protein and the DNA contents of fibroblasts were evaluated with 7-AAD using flow cytometry. Representatives of FACS plots of fibroblasts from a healthy individual (control) and an individual with PDP (PDP) are shown (A). The percentages of fibroblasts in S-G2 phase in triplicated wells are expressed as the mean ± SD (n=3). The student's *t* test was performed between the indicated groups and an asterisk indicates $P < 0.05$ (B). (C) The actin bundle formation of cultured fibroblasts from a healthy individual (control) and an individual with PDP were examined by staining with alexa 488-labeled phalloidin antibody 5 days after the 4th passage. Upper panels, low magnification (×10); lower panels, high magnification (×40) (adapted from ref. (Kabashima et al., 2010)).

Enhanced Actin Bundle Formation of Fibroblasts in PDP

Wnt signaling is also known to induce cell motility and cytoskeletal rearrangement of NIH3T3, a fibroblast cell line (Kim and Choi, 2007). Therefore, we examined the actin bundle formation of cultured fibroblasts with phalloidin staining 5 days after the 4th passage. Fluorescent microscopy showed that the actin bundle formation of PDP fibroblasts is promoted in PDP, as the bundles were thicker and denser than those of control fibroblasts (Figure 3. C).

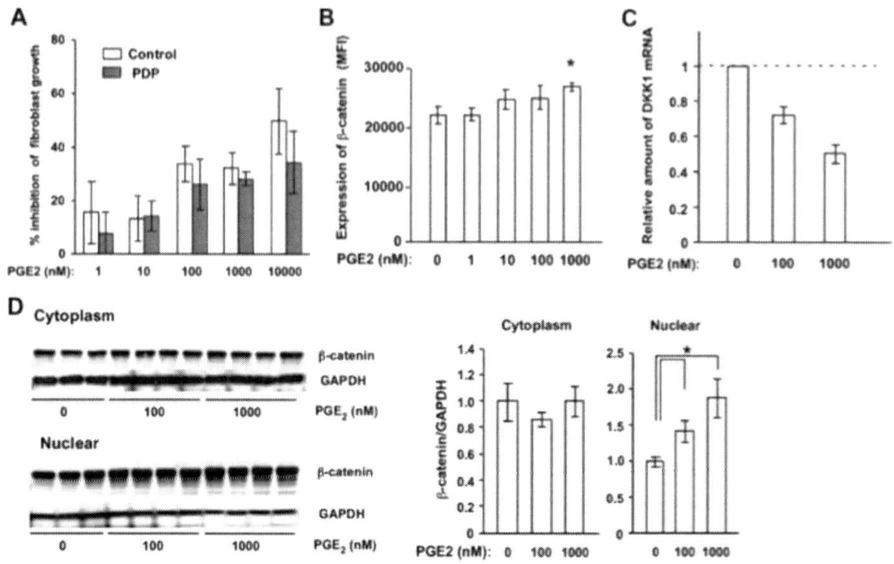

Figure 4. Effect of PGE_2 on fibroblasts. (A) The % inhibition of the number of fibroblasts from a healthy donor and a PDP patient by the addition of PGE_2 was evaluated as (Number of fibroblasts without PGE_2 − Number of fibroblasts with PGE_2)/Number of fibroblasts without $PGE_2 \times 100$. The growth inhibitory effect of PGE_2 is dose-dependent and comparable between these two groups. The values are expressed as the mean ± SD (n=3) and are representative of two independent experiments. (B, C) The effects of PGE_2 on β-catenin expression and *DKK1* mRNA levels in fibroblasts were evaluated. The mean fluorescent intensity (MFI) of β-catenin (B) and *DKK1* mRNA (C) in fibroblasts after exposure to PGE_2 is shown. The amount of *DKK1* mRNA relative to *GAPDH* mRNA without the addition of PGE_2 is regarded as one. The values are expressed as the mean ± SD (n=3) and an asterisk indicates $P < 0.05$. (D) Cytoplasm- (upper panel) and nuclear- (lower panel) protein samples from fibroblasts treated with or without 0, 100, and 1000 nM PGE_2 for 4 days were used to determine the effect of PGE_2 on β-catenin expression. The values are expressed as the mean ± SD (n=3-4) and an asterisk indicates $P < 0.05$ (adapted from ref. (Kabashima et al., 2010)).

Effect of PGE_2 on Fibroblasts

It was recently reported that PDP is induced by elevated PGE_2 due to mutations in the *HPGD and/or SLCO2A1* genes (Uppal et al., 2008). If this PGE_2 alteration also affects pachydermia, PGE_2 is expected to enhance fibroblast function and proliferation. The addition of PGE_2 into the cultured medium of fibroblasts decreased the number of dermal fibroblasts from healthy donors in a dose-dependent manner as reported previously (Fine and Goldstein, 1987; Korn et al., 1980) (Figure 4. A). A similar effect was observed when PGE_2 was added to the culture medium of fibroblasts from the PDP patient (case 2). To examine whether PGE_2 affects Wnt signaling in fibroblasts, we measured the amount of β-catenin in fibroblasts after exposure to PGE_2 by flow cytometry, and found that β-catenin was significantly increased in fibroblasts by the addition of PGE_2 at a dose of 1000 nM (Figure 4. B). In addition, the mRNA expression level of *DKK1* was significantly decreased by the addition of PGE_2 at a dose of 100 and 1000 nM (Figure 4. C). Moreover, to determine the effect of PGE_2 on β-catenin expression, cytoplasm- and nuclear- protein samples were prepared from fibroblasts treated with or without 0, 100, and 1000 nM PGE_2 in the presence of 10 μM indomethacin for 4 days. In the cytoplasm, β-catenin expression was unchanged irrespective of the addition of PGE_2. However, β-catenin expression in the nuclei was significantly increased by the treatment with 100 and 1000 nM PGE_2 (Figure 4. D). These results suggest that PGE_2 signaling increases nuclear β-catenin in fibroblasts.

DKK1 siRNA Enhances Ear Thickness in Mice

Finally, we used mice to pursue direct evidence for DKK1 involvement in pachydermia. We injected a solution of mouse *DKK1* siRNA or control siRNA intradermally into the ears of mice 4 times every 7 days. Quantitative RT-PCR analysis revealed that this procedure successfully suppressed the expression of *DKK1* mRNA in the skin by about 60% (Figure 5. A).

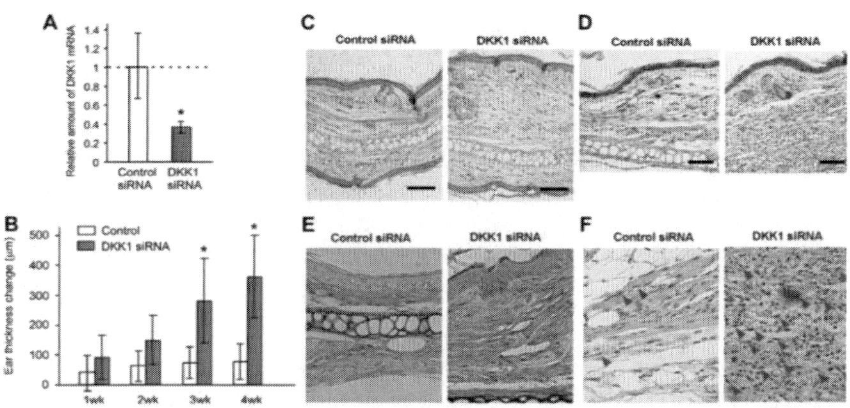

Figure 5. The effect of mouse *DKK1* siRNA on pachydermia. (A) Mouse *DKK1* siRNA and control-scrambled siRNA (control siRNA) solutions were injected intradermally into the ears of mice 4 times every 7 days. The expression of *DKK1* mRNA in the skin 1 week after the last siRNA injection was evaluated by quantitative RT-PCR analysis. The mean value of *DKK1* mRNA relative to *GAPDH* mRNA treated with scrambled siRNA is regarded as one (n=5). (B) The ear thickness was measured every week before each injection and one week after the last injection. *Columns show the mean ± SD* (n=8, each group) *from* two independent experiments. The student's *t* test was performed between the indicated groups and an asterisk indicates $P < 0.05$. (C-D) The skin from the ears one week after the last siRNA injection was fixed and stained with hematoxylin and eosin (C). In addition, skin sections were stained with anti-β-catenin antibody by immunohistochemistry. Enhanced β-catenin expression is visible diffusely in the dermis of the skin treated with *DKK1* siRNA (D). Scale bars = 100 μm. (E) The samples stained with Elastica van Gieson shows the thick collagen and elastic fibers in the dermis of mice treated with DKK1 siRNA. (F) The samples are stained with PCNA (adapted from ref. (Kabashima et al., 2010)).

The ear thickness was measured every week before each injection and one week after the last injection. The ear thickness was significantly augmented by the application of *DKK1* siRNA (Figure 5. B). The histological findings showed that the dermis was thickened with increased fibroblasts (Figure 5. C). Consistent with these findings, enhanced β-catenin expression was observed diffusely in the dermis treated with *DKK1 si*RNA. (Figure 5. D).

Conclusion

We showed that Wnt/DKK1 played a key role in the development of pachydermia in several aspects (Figure 6). Firstly, proliferation of fibroblasts from the PDP patients was promoted with a higher ratio in the cell cycle than compared to normal fibroblasts, and human recombinant DKK1 protein decreased their proliferation. Secondly, the expression levels of DKK1 mRNA in PDP fibroblasts and DKK1 protein in PDP skin were lower than those in healthy controls. Thirdly, β-catenin intensity in the skin from PDP was pronounced by immunohistochemistry. Finally, application of mouse *DKK1* siRNA increased the thickness of the skin in accordance with the elevated β-catenin levels. These results suggest that enhanced Wnt signaling is related to the development of pachydermia.

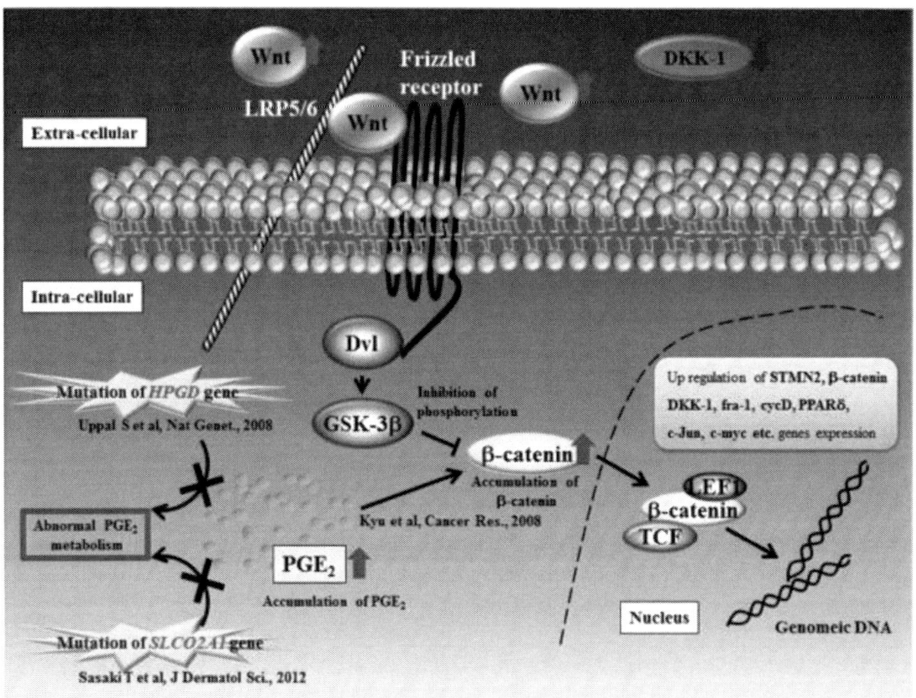

Figure 6. Schema of the pathomechanism of pachydermoperiostosis with the mandatory role of dermal fibroblasts.

Pachydermia is one of the clinical manifestations of the complete form of PDP, which involves both skin and bone. For example, BMP, TGF-β, and Wnt families are the possible molecules responsible for the changes in both organs. There are several congenital diseases related to both organs, such as basal cell nevus syndrome, synovitis acne pustulosis hyperostosis ostitis syndrome, Klippel-Trenaunay-Weber syndrome, and Buschke-Ollendorff syndrome (Ehrig and Cockerell, 2003; Kahn and Chamot, 1992; Oduber et al., 2008). Bushchke-Ollendorff syndrome, in which osteopoikilosis is associated with connective tissue nevi, is particularly of note, since mutations in *LEMD3*, a gene implicated in BMP signaling, are candidates for its pathogenesis (Hellemans et al., 2004). However, we could not detect a significant difference in mRNA expression for *BMP* or *TGF-β* families between PDP and control fibroblasts by DNA microarray analysis.

Recently, the incomplete form of PDP was attributed to elevated PGE_2 due to the mutation of *HPGD*. The skeletal phenotype of PDP, particularly clubbing and periostosis, can clearly be explained by elevated PGE_2, since PGE_2 is known to stimulate the activity of both osteoclasts and osteoblasts (Raisz et al., 1993), leading to bone deposition (periostosis) and resorption (acro-osteolysis), respectively. However, we could not detect *HPGD* gene mutations in our patients. Intriguingly, *SLCO2A1* gene mutations were identified in complete or incomplete form of PDP patients recently (Diggle et al., 2012; Sasaki et al., 2012). Clinical data and genetic analyses showed that *SLCO2A1* genotypes in PDP patients were closely associated with serum PGE_2 levels, suggesting that *SLCO2A1* mutations contribute to the severity of clinical phenotypes in PDP (Busch et al., 2012; Diggle et al., 2012; Sasaki et al., 2012; Seifert et al., 2012; Zhang et al., 2012). In addition, although the serum PGE_2 level of patients with the incomplete type PDP patients were within the normal range, we found mild sebaceous gland hyperplasia. These results suggested that the mild *SLCO2A1* mutation found in the patients with the incomplete form of PDP could alter the PGE_2 level in affected skin, although serum PEG_2 level would be within the normal range (Sasaki et al., 2012).

Here we focused on Wnt signaling in the development of pachydermia. Fibroblasts from PDP skin and bone marrow-derived fibroblasts of PDP patients are known to grow faster than those of healthy donors (Fontenay-Roupie et al., 1995; Matucci-Cerinic et al., 1992). The transfection of *DKK1* into cultured mouse fibroblasts, NIH3T3, blocked WNT2-induced cell growth and the WNT2-induced increase in uncomplexed β-catenin (Fedi et al., 1999). WNT3a induced motility and cytoskeletal rearrangement of NIH3T3 cells (Kim and Choi, 2007). These previous reports suggest that enhanced Wnt/β-catenin signaling promotes fibroblast proliferation and cytoskeletal rearrangement. In fact, we found that the frequency of PDP fibroblasts in cycle was increased, and that actin bundle formation was more pronounced in PDP fibroblasts. Moreover, the addition of human recombinant *DKK1* consistently suppressed the fibroblast proliferation.

The source of DKK1 and how it works in the skin are issues that remain to be clarified. According to our immunohistochemical analysis, the major source of DKK1 in the skin seems to be fibroblasts, because the DKK1 expression in fibroblasts was low in PDP. Since DKK1 is a secreted antagonist and may affect bystander cells in the vicinity of fibroblasts, the dysregulated production of DKK1 possibly modulates the functions of not only fibroblasts but also other cells, such as keratinocytes and melanocytes. It was reported that high *DKK1* expression by dermal fibroblasts in the palms and soles inhibited the function of melanocytes via suppression of β-catenin and MITF, and enhanced keratinocyte proliferation (Yamaguchi et al., 2004; Yamaguchi et al., 2008; Yamaguchi et al., 2007). Mice with an overexpression of *DKK1* in skin consistently lacked formation of appendages, such as hair follicles, and the mice had no skin pigmentation on the trunk (Andl et al., 2002).

The role of DKK1 has been more extensively studied in bone than in the skin. DKK1 is known to inhibit osteoblast differentiation, and the overproduction of DKK1 was noted in osteolytic bone lesions of patients with multiple myeloma (Qiang et al., 2003). The elevated *DKK1* levels in bone marrow plasma and peripheral blood from the patients were correlated with the presence of focal bone lesions. Recombinant human *DKK1* inhibited the differentiation of osteoblast precursor cells *in vitro* (Morvan et al., 2006; Tian et al., 2003). These previous observations could explain the periostosis in PDP possibly secondary to decreased *DKK1* expression. Since fibroblasts and osteoblasts are derived from mesenchymal origin, they seem to share in common the mechanism of differentiation and proliferation. Although we did not address the relationship between DKK1 and the skeletal phenotype in PDP, it would be of interest to analyze the function of osteoblasts in PDP.

The next question is how Wnt signaling is enhanced. One possibility provided by our present study is the suppression of DKK1 expression in fibroblasts. The mechanism by which DKK1 is downregulated in PDP remains to be elucidated. It can be hypothesized that there is a mutation in *DKK1* or molecules controlling *DKK1* expression, such as TCF-4. However, no mutation was detected in either exon of DKK1 or TCF-4 genes. Therefore, in the present study, we could not determine the genetic mechanism responsible for the complete form of PDP and/or pachydermia. Given the defect in PDP appears to altered expression of DKK1, it will be of interest in future studies to analyze the regulatory regions of DKK1, especially around the TCF binding sites, an issue which remains to be clarified.

On the other hand, the Wnt/β-catenin pathway is known to increase *DKK1* mRNA and protein, thus initiating a negative feedback loop (Gonzalez-Sancho et al., 2005). It can be hypothesized that this negative feedback regulation might be dysregulated in PDP. Moreover, due to this negative feedback system, DKK1 can work as a tumor suppressor gene in some types of neoplasia (Gonzalez-Sancho et al., 2005; Polakis, 2000). Hypertrophic osteoarthritis is occasionally induced by a variety of thoracoabdominal, sometimes malignant, conditions. The relationship between decreased DKK1 expression and secondary hypertrophic osteoarthritis in association with malignancy may be an interesting issue to pursue.

It still remains unclear whether PDP in our cases could be attributed to the mutation in HPGD or not. Of note is that our cases were diagnosed as the complete form of PDP including pachydermia and adolescent onset, but that the cases with HPGD mutation had the incomplete form of PDP without pachydermia and with early onset (within the first year of their lives).

The onset of the PDP is bimodal. The first peak is during the first year of the life and the second at the age of 15 years (Matucci-Cerinic et al., 1991; Rimoin, 1965). Therefore, the pathogenesis of PDP might be subdivided into at least two groups. However, further clinical studies in combination with HPGD mutation analysis will be required to clarify this.

In PDP, clinical cutaneous manifestations include pachydermia, seborrhea, and velvet colored skin. At present, we could not show direct evidence that all the phenotypes of PDP were induced by enhanced Wnt signaling secondary to the suppressed expression of DKK1. In addition, the number of cases in our study was limited. However, our findings together with those of previous studies suggest that the Wnt signaling pathway was promoted in accordance with decreased DKK1 expression, leading to increased fibroblast proliferation, enhanced pigmentation of the skin, and adnexal hyperplasia.

References

Andl, T., Reddy, S. T., Gaddapara, T., and Millar, S. E. (2002). WNT signals are required for the initiation of hair follicle development. *Dev. Cell 2*, 643-653.

Botchkarev, V. A., and Sharov, A. A. (2004). BMP signaling in the control of skin development and hair follicle growth. *Differentiation 72*, 512-526.

Busch, J., Frank, V., Bachmann, N., Otsuka, A., Oji, V., Metze, D., Shah, K., Danda, S., Watzer, B., Traupe, H., *et al.* (2012). Mutations in the prostaglandin transporter SLCO2A1 cause primary hypertrophic osteoarthropathy with digital clubbing. *The Journal of investigative dermatology, 132*, 2473-2476.

Castori, M., Sinibaldi, L., Mingarelli, R., Lachman, R. S., Rimoin, D. L., and Dallapiccola, B. (2005). Pachydermoperiostosis: an update. *Clin. Genet. 68*, 477-486.

Diggle, C. P., Parry, D. A., Logan, C. V., Laissue, P., Rivera, C., Restrepo, C. M., Fonseca, D. J., Morgan, J. E., Allanore, Y., Fontenay, M., *et al.* (2012). Prostaglandin transporter mutations cause pachydermoperiostosis with myelofibrosis. *Human mutation 33*, 1175-1181.

Ehrig, T., and Cockerell, C. J. (2003). Buschke-Ollendorff syndrome: report of a case and interpretation of the clinical phenotype as a type 2 segmental manifestation of an autosomal dominant skin disease. *J. Am. Acad. Dermatol. 49*, 1163-1166.

Fedi, P., Bafico, A., Nieto Soria, A., Burgess, W. H., Miki, T., Bottaro, D. P., Kraus, M. H., and Aaronson, S. A. (1999). Isolation and biochemical characterization of the human Dkk-1 homologue, a novel inhibitor of mammalian Wnt signaling. *J. Biol. Chem. 274*, 19465-19472.

Fine, A., and Goldstein, R. H. (1987). The effect of PGE2 on the activation of quiescent lung fibroblasts. *Prostaglandins 33*, 903-913.

Fontenay-Roupie, M., Dupuy, E., Berrou, E., Tobelem, G., and Bryckaert, M. (1995). Increased proliferation of bone marrow-derived fibroblasts in primitive hypertrophic osteoarthropathy with severe myelofibrosis. *Blood 85*, 3229-3238.

Glinka, A., Wu, W., Delius, H., Monaghan, A. P., Blumenstock, C., and Niehrs, C. (1998). Dickkopf-1 is a member of a new family of secreted proteins and functions in head induction. *Nature 391*, 357-362.

Gonzalez-Sancho, J. M., Aguilera, O., Garcia, J. M., Pendas-Franco, N., Pena, C., Cal, S., Garcia de Herreros, A., Bonilla, F., and Munoz, A. (2005). The Wnt antagonist DICKKOPF-1 gene is a downstream target of beta-catenin/TCF and is downregulated in human colon cancer. *Oncogene 24*, 1098-1103.

Grotewold, L., and Ruther, U. (2002). The Wnt antagonist Dickkopf-1 is regulated by Bmp signaling and c-Jun and modulates programmed cell death. *Embo J. 21*, 966-975.

Hellemans, J., Preobrazhenska, O., Willaert, A., Debeer, P., Verdonk, P. C., Costa, T., Janssens, K., Menten, B., Van Roy, N., Vermeulen, S. J., *et al.* (2004). Loss-of-function mutations in LEMD3 result in osteopoikilosis, Buschke-Ollendorff syndrome and melorheostosis. *Nat. Genet. 36*, 1213-1218.

Jajic, I., Jajic, Z., and Grazio, S. (2001). Minor but important symptoms and signs in primary hypertrophic osteoarthropathy. *Clin. Exp. Rheumatol. 19*, 357-358.

Kabashima, K., Banks, T. A., Ansel, K. M., Lu, T. T., Ware, C. F., and Cyster, J. G. (2005). Intrinsic lymphotoxin-beta receptor requirement for homeostasis of lymphoid tissue dendritic cells. *Immunity 22*, 439-450.

Kabashima, K., Haynes, N. M., Xu, Y., Nutt, S. L., Allende, M. L., Proia, R. L., and Cyster, J. G. (2006). Plasma cell S1P1 expression determines secondary lymphoid organ retention versus bone marrow tropism. *J. Exp. Med. 203*, 2683-2690.

Kabashima, K., Sakabe, J., Yoshiki, R., Tabata, Y., Kohno, K., and Tokura, Y. (2010). Involvement of Wnt signaling in dermal fibroblasts. *The American journal of pathology 176*, 721-732.

Kahn, M. F., and Chamot, A. M. (1992). SAPHO syndrome. *Rheum. Dis. Clin. North Am. 18*, 225-246.

Kawano, Y., and Kypta, R. (2003). Secreted antagonists of the Wnt signalling pathway. *J. Cell Sci. 116*, 2627-2634.

Kim, S. E., and Choi, K. Y. (2007). EGF receptor is involved in WNT3a-mediated proliferation and motility of NIH3T3 cells via ERK pathway activation. *Cell Signal 19*, 1554-1564.

Korn, J. H., Halushka, P. V., and LeRoy, E. C. (1980). Mononuclear cell modulation of connective tissue function: suppression of fibroblast growth by stimulation of endogenous prostaglandin production. *J. Clin. Invest. 65*, 543-554.

Kornak, U., and Mundlos, S. (2003). Genetic disorders of the skeleton: a developmental approach. *Am. J. Hum. Genet. 73*, 447-474.

Kushibiki, T., Matsumoto, K., Nakamura, T., and Tabata, Y. (2004). Suppression of tumor metastasis by NK4 plasmid DNA released from cationized gelatin. *Gene Ther. 11*, 1205-1214.

Livak, K. J., and Schmittgen, T. D. (2001). Analysis of relative gene expression data using real-time quantitative PCR and the 2(-Delta Delta C(T)) *Method. Methods 25*, 402-408.

Maekawa, M., Yamamoto, T., Tanoue, T., Yuasa, Y., Chisaka, O., and Nishida, E. (2005). Requirement of the MAP kinase signaling pathways for mouse preimplantation development. *Development 132*, 1773-1783.

Mao, B., Wu, W., Davidson, G., Marhold, J., Li, M., Mechler, B. M., Delius, H., Hoppe, D., Stannek, P., Walter, C., *et al.* (2002). Kremen proteins are Dickkopf receptors that regulate Wnt/beta-catenin signalling. *Nature 417*, 664-667.

Mao, B., Wu, W., Li, Y., Hoppe, D., Stannek, P., Glinka, A., and Niehrs, C. (2001). LDL-receptor-related protein 6 is a receptor for Dickkopf proteins. *Nature 411*, 321-325.

Masuda, K., Moriwaki, S., Takigawa, M., Furukawa, F., Higashishiba, T., and Fukamizu, H. (2000). A case of pachydermoperiostosis. Rinsho Hifuka (Japanese) *54*, 398-401.

Matucci-Cerinic, M., Lotti, T., Jajic, I., Pignone, A., Bussani, C., and Cagnoni, M. (1991). The clinical spectrum of pachydermoperiostosis (primary hypertrophic osteoarthropathy). *Medicine* (Baltimore) *70*, 208-214.

Matucci-Cerinic, M., Sacerdoti, L., Perrone, C., Carossino, A., Cagnoni, M. L., Jajic, I., and Lotti, T. (1992). Pachydermoperiostosis (primary hypertrophic osteoarthropathy): in vitro evidence for abnormal fibroblast proliferation. *Clin. Exp. Rheumatol. 10 Suppl 7*, 57-60.

Morvan, F., Boulukos, K., Clement-Lacroix, P., Roman Roman, S., Suc-Royer, I., Vayssiere, B., Ammann, P., Martin, P., Pinho, S., Pognonec, P., *et al.* (2006). Deletion of a single allele of the Dkk1 gene leads to an increase in bone formation and bone mass. *J. Bone Miner. Res. 21*, 934-945.

Oduber, C. E., van der Horst, C. M., and Hennekam, R. C. (2008). Klippel-Trenaunay syndrome: diagnostic criteria and hypothesis on etiology. *Ann. Plast. Surg. 60*, 217-223.

Polakis, P. (2000). Wnt signaling and cancer. *Genes Dev. 14*, 1837-1851.

Qiang, Y. W., Endo, Y., Rubin, J. S., and Rudikoff, S. (2003). Wnt signaling in B-cell neoplasia. *Oncogene 22*, 1536-1545.

Raisz, L. G., Pilbeam, C. C., and Fall, P. M. (1993). Prostaglandins: mechanisms of action and regulation of production in bone. *Osteoporos. Int. 3 Suppl 1*, 136-140.

Rimoin, D. L. (1965). Pachydermoperiostosis (Idiopathic Clubbing And Periostosis): Genetic And Physiologic Considerations. *N. Engl. J. Med. 272*, 923-931.

Sasaki, T., Niizeki, H., Shimizu, A., Shiohama, A., Hirakiyama, A., Okuyama, T., Seki, A., Kabashima, K., Otsuka, A., Ishiko, A., *et al.* (2012). Identification of mutations in the prostaglandin transporter gene SLCO2A1 and its phenotype-genotype correlation in Japanese patients with pachydermoperiostosis. *Journal of dermatological science 68*, 36-44.

Seifert, W., Kuhnisch, J., Tuysuz, B., Specker, C., Brouwers, A., and Horn, D. (2012). Mutations in the prostaglandin transporter encoding gene SLCO2A1 cause primary hypertrophic osteoarthropathy and isolated digital clubbing. *Human mutation 33*, 660-664.

Shawarby, K., and Ibrahim, M. S. (1962). Pachydermoperiostosis. A review of literature and report on four cases. *Br. Med. J. 1*, 763-766.

Tian, E., Zhan, F., Walker, R., Rasmussen, E., Ma, Y., Barlogie, B., and Shaughnessy, J. D., Jr. (2003). The role of the Wnt-signaling antagonist DKK1 in the development of osteolytic lesions in multiple myeloma. *N. Engl. J. Med. 349*, 2483-2494.

Touraine, A., Solente, G., and Gole, A. (1935). Un syndrome osteodermopathique: la pachydermie plicaturee avec pachyperiostose des extremites. Presse Med *43*, 1820-1824.

Uppal, S., Diggle, C. P., Carr, I. M., Fishwick, C. W., Ahmed, M., Ibrahim, G. H., Helliwell, P. S., Latos-Bielenska, A., Phillips, S. E., Markham, A. F., *et al.* (2008). Mutations in 15-hydroxyprostaglandin dehydrogenase cause primary hypertrophic osteoarthropathy. *Nat. Genet. 40*, 789-793.

Vogl, A., and Goldfischer, S. (1962). Pachydermoperiostosis. Primary or idiopathic hypertrophic osteoarthropathy. *Am. J. Med. 33*, 166-187.

Yamaguchi, Y., Itami, S., Watabe, H., Yasumoto, K., Abdel-Malek, Z. A., Kubo, T., Rouzaud, F., Tanemura, A., Yoshikawa, K., and Hearing, V. J. (2004). Mesenchymal-epithelial interactions in the skin: increased expression of dickkopf1 by palmoplantar fibroblasts inhibits melanocyte growth and differentiation. *J. Cell Biol. 165*, 275-285.

Yamaguchi, Y., Passeron, T., Hoashi, T., Watabe, H., Rouzaud, F., Yasumoto, K., Hara, T., Tohyama, C., Katayama, I., Miki, T., and Hearing, V. J. (2008). Dickkopf 1 (DKK1) regulates skin pigmentation and thickness by affecting Wnt/beta-catenin signaling in keratinocytes. *FASEB J. 22*, 1009-1020.

Yamaguchi, Y., Passeron, T., Watabe, H., Yasumoto, K., Rouzaud, F., Hoashi, T., and Hearing, V. J. (2007). The effects of dickkopf 1 on gene expression and Wnt signaling by melanocytes: mechanisms underlying its suppression of melanocyte function and proliferation. *J. Invest. Dermatol. 127*, 1217-1225.

Zhang, Z., Xia, W., He, J., Ke, Y., Yue, H., Wang, C., Zhang, H., Gu, J., Hu, W., Fu, W., *et al.* (2012). Exome sequencing identifies SLCO2A1 mutations as a cause of primary hypertrophic osteoarthropathy. *American journal of human genetics 90*, 125-132.

Zorn, A. M. (2001). Wnt signalling: antagonistic Dickkopfs. *Curr. Biol. 11*, R592-595.

In: Dermal Fibroblasts
Editor: Xiaowen Bai

ISBN: 978-1-62618-799-3

Chapter 8

Stem Cell-Like Characteristics of Dermal Fibroblasts: Proliferation, Antigen Profiles, and Differentiation Capacity

Xiaowen Bai*
Department of Anesthesiology, Medical College of Wisconsin, Milwaukee, WI, US

Abstract

Fibroblasts are plastic adherent mesenchymal cells that play a significant role during tissue development, maintenance, and repair. Meshenchymal stem cells (MSCs) are adherent fibroblast-like cells and are initially isolated from adipose tissue, bone marrow, and skeketal muscle. MSCs are characterized by their extensive proliferative capacity and the ability to differentiate into other cell lineages such as adipocytes and osteoblasts. In contrast to MSCs, fibroblasts are widely considered to be terminally differentiated cells that are primarily responsible for the synthesis and remodeling of extracellular matrix in tissues and do not convert into other types of cells. Although fibroblasts and MSCs have been well studied, differences between these two cell types are not fully understood. Growing experimental evidence demonstrate that skin fibroblasts share similar characteristics with MSCs (e.g., antigen and gene profiles, colony-forming ability, and differentiation potential) as described as follows: (1) fibroblasts and MSCs exhibit a similar spindle-like morphology and express similar antigen profiles: positive for MSC markers (CD44, CD73, and CD105) and fibroblastic markers (collagen I, heat shock protein 47, vimentin, fibroblast surface protein, and α smooth muscle actin), and negative for endothelial cell marker (CD31) and hematopoietic lineage markers (CD14 and CD45); (2) these two types of cells secrete angiogenic growth factors (e.g., VEGF); and (3) both fibroblasts and MSCs show the capacity to form colonies and differentiate into adipocytes, osteoblasts, and chondrocytes *in vitro*. *In vivo* animal studies also indicate the

* Corresponding Author address: 8701 Watertown Plank, Milwaukee, WI 53226, USA. Email: baixw01@yahoo.com.

differentiation potential of fibroblasts. Collectively, these findings suggest that (1) fibroblast preparations from the skin consist of a significant number of stem cells with differentiation potential; (2) skin fibroblasts might play an important role in cutaneous wound healing; and (3) skin fibroblasts can be a promising alternative autologous cell source for tissue regeneration because of their easy access. Importantly, these findings shed new light on the commonly used definition of fibroblasts and provide novel insights into many past studies performed with fibroblast preparations since the reported effects might be greatly attributable to stem cell content within skin fibroblast preparations. This chapter will review the stem cell-like characteristics of dermal fibroblasts.

Keywords: Fibroblasts, adipose tissue-derived stem cells, differentiation

Introduction

Skin is the largest organ of the human body and has three layers: epidermis, dermis, and a fat layer. The key type of cells in dermis are fibroblasts. Fibroblasts are plastic adherent mesenchymal cells. They are widely considered to be terminally differentiated cells and do not convert into other types of cells. Fibroblasts are primarily responsible for the synthesis and remodeling of extracellular matrix in tissues and play a significant role during tissue development, maintenance, and repair (Flavell et al., 2008).

Mesenchymal stem cells (MSCs) are currently defined as plastic adherent fibroblast-like cells with extensive proliferative capacity and differentiation potential. They have been identified in many organs and tissues, including bone marrow (Pittenger et al., 1999), adipose tissue (Bai et al., 2007), skin (Richardson et al., 2005), muscle, and periosteum (Kisiel et al., 2012). We and others have shown that adipose tissue-derived stem cells (ASCs) improved cardiac function and promoted cutaneous wound healing (Altman et al., 2008a) (Altman et al., 2008b; Bai et al., 2010). Bone marrow-derived stem cells (BMSCs) also exhibited the benefits in tissue regeneration (e.g., fracture healing) (Park et al., 2012).

Traditionally, differentiation capacity has been used a line of demarcation to distinguish fibroblasts from MSCs. Therefore, fibroblasts are routinely used as a negative control for evaluation of cell differentiation capacity in many studies (Brendel et al., 2005). However, recent experimental evidence demonstrated that dermal fibroblasts were able to differentiate into various lineages such as adipocytes and chondrocytes *in vitro* and *in vivo*. The fibroblasts isolated from bronchial and uterine tissues also showed the similar differentiation potential (Kim et al., 2008; Strakova et al., 2008).

The differentiation potential of fibroblasts leads to obscure the line of demarcation to distinguish fibroblasts from stem cells (Alt et al., 2011; Chen et al., 2007) and raises the following questions: how are fibroblasts defined, how do fibroblasts differ from MSCs, and are the beneficial effects obtained from so called fibroblasts (that have been described nearly for a decade) mediated by stem cells or by fibroblasts? Answering these questions is particularly important for understanding the biology of fibroblasts and stem cells, explaining the related experimental results, and evaluating the effect of fibroblasts on tissue regeneration. Here, we are going to summarize the recent findings regarding the identification of stem cell characteristics of fibroblasts such as morphology, proliferation potential, colony-forming ability, antigen and gene profiles, growth factor secretion, differentiation potential, and the relationship between fibroblasts and stem cells.

Fibroblasts and Stem Cells Have Similar Shape and Proliferation Potential

The data from our lab showed that human dermal fibroblasts and ASCs exhibited similar spindle-like morphology (Fig. 1) and proliferative potential *in vitro*. The cell numbers of both dermal fibroblasts and ASCs expanded to about 600,000 from 250,000 in 24-cm^2 flasks 10 days after seeding. Cell proliferation capacity was further analyzed by the following colony-forming unit (CFU) assay. Fibroblasts and ASCs were plated at a density of 500 cells in a 60-mm culture dish (17.69 cells/cm^2). After 14 days' culture, cells were stained with saturated methyl violet, indicating the colonies. The colonies containing more than 50 cells were then counted. The colony-forming efficiency of cells was calculated by dividing the number of colonies formed per dish by the initial number of cells (500) seeded per dish. The colony-forming efficiency of fibrblasts (5 %) was similar to that of ASCs (4 %) (Alt et al., 2011). These data indicate that there are no significant differences in the proliferation capability between dermal fibroblasts and ASCs.

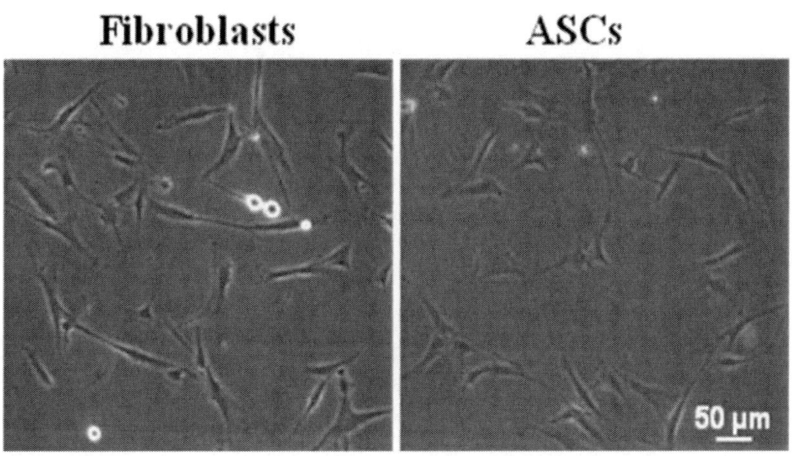

Figure 1. Morphology of human dermal fibroblasts and adipose tissue-derived stem cells (ASCs) *in vitro*. These two types of cells exhibit similar spindle-like shape (Alt et al., 2011).

Fibroblasts and Stem Cells Express Both Stem Cell and Fibroblast Markers

Stem Cell Marker Expression

Flow cytometry assay indicated that dermal fibroblasts were positive for MSC markers (CD44, 96 ± 2 %; CD73, 99 ± 1%; CD90, 97 ± 3 %; CD105, 100%). Both hematopoietic cell markers (CD14, 0.19 ± 0.39%; CD45, 0.12 ± 0.34%) and endothelial cell marker (CD31, 0.23 ± 0.25%) were absent in fibrobasts. This result precludes the existence of hematopoietic cells and endothelial cells in the cuture (Etheridge et al., 2004). Cell surface proteins expressed in ASCs were highly comparable to those in dermal fibroblasts. ASCs were also positive for

CD44 (92 ± 5 %), CD73 (98 ± 2), and CD105 (93 ± 7 %) and negative for CD14 (0.11 ± 0.11%), CD45 (0.30 ± 0.37%), and CD31 (0.11 ± 0.33%) (Alt et al., 2011). The expression pattern of cell surface markers in fibroblasts and ASCs were consistent with the pattern of cell surface markers in other tissue-derived MSCs (e.g., BMSCs) (Izadpanah et al., 2006). In addition, immunofluorescence staining analysis displayed OCT4 expression in both fibroblasts and ASCs. OCT4, a transcription factor, is highly associated with pluripotency of stem cells (Buitrago and Roop, 2007; Choo et al., 2004).

Stem Cell-Associated Gene Expression

The expression of 90 human stem cell-related genes in human dermal fibroblasts and ASCs were analyzed by real-time PCR. These genes include embryonic stem cell markers (e.g., OCT4, KLF4, NANOG, LIN28, FGF4, REST, SALL2, and CER1), mesoderm markers (e.g., FN1, SFRP2 and NPPA), and endoderm markers (e.g., SERPINA1). There was a similar gene expression pattern between these two types of cells. Specificially, seventy genes were commonly expressed in fibroblasts and ASCs. (Alt et al., 2011).

Fibroblast Marker Expression

Collagen I, HSP47 (heat shock protein 47), vimentin, and FSP (fibroblast surface protein) are commonly regarded as the markers of fibroblasts (Kalluri and Zeisberg, 2006; Strutz et al., 1995). αSMA (αlpha smooth muscle actin) is a myofibroblasts marker (McAnulty, 2007). Fibroblasts and ASCs were all positive for these fibroblast markers (Alt et al., 2011).

Collagen I is one of the primary collagens in skin. It provides structure, strength, and integrity of skin and is involved in stimulating cell proliferation and differentiation, guiding cell migration, and modulating cellular responses (Midwood et al., 2004). HSP47 is a collagen-specific chaperone that specifically recognizes the correctly folded triple helical region of collagen in the endoplasmic reticulum and is involved in the folding and maturation of collagen. HSP47 knockout embryos died just after 10.5 days due to the absence of functional collagen (Kubota and Nagata, 2004). Vimentin, belongs to type III intermediate filaments, plays a fundamental role in maintaining mesenchymal cell shape and integrity, and stabilizing cytoskeletal interactions and cell substrate adhesion (Goldman et al., 1996). The expression of fibroblast antigens in fibroblasts and ASCs indicates that these two types of cells might have common function in collagen secretion for tissue maintenance and wound healing (Alt et al., 2011).

Fibroblasts Exhibit Differentiation Potential

To evaluate the differentiation capacity of human dermal fibroblasts and ASCs, cells were cultured in adipogenic, osteogenic, and chondrogenic induction media for 3 weeks. Both fibroblasts and ASCs cultured in adipogenic media displayed multiple intracellular bright white oil droplets. These oil droplets showed red vesicles when stained with Oil Red O. There were more Oil Red O-positive adipocytes in ASC culture (60 ± 8 %) than in fibroblast culture (44 ± 6 %). Fibroblasts and ASCs cultured in osteogenic media showed black regions within the monolayer, indicating calcium deposits from the osteoblasts differentiated ASCs. There were no obvious differences in osteogenic capacity between fibroblasts and ASCs as determined by measuring the intensity of calcium deposition following Alizarin Red S staining. Cell pellet sections from fibroblasts and ASCs cultured for 3 weeks in chondrogenic media showed brown signals after immunohistochemical staining. The brown signals represent the expression of collagen II (a specific marker of chondrocytes). There were no obvious differences of chondrogenic potency between fibroblasts and ASCs (Fig. 2) (Alt et al., 2011; Lorenz et al., 2008). Fibroblasts were also able to differentiate into other cell types including neurons, skeletal myocytes, and insulin-producing cells (Osonoi M, 2011). Collectively, these findings indicate that human dermal fibroblasts have more plasticity than what has generally been considered and may have the potential for tissue regeneration.

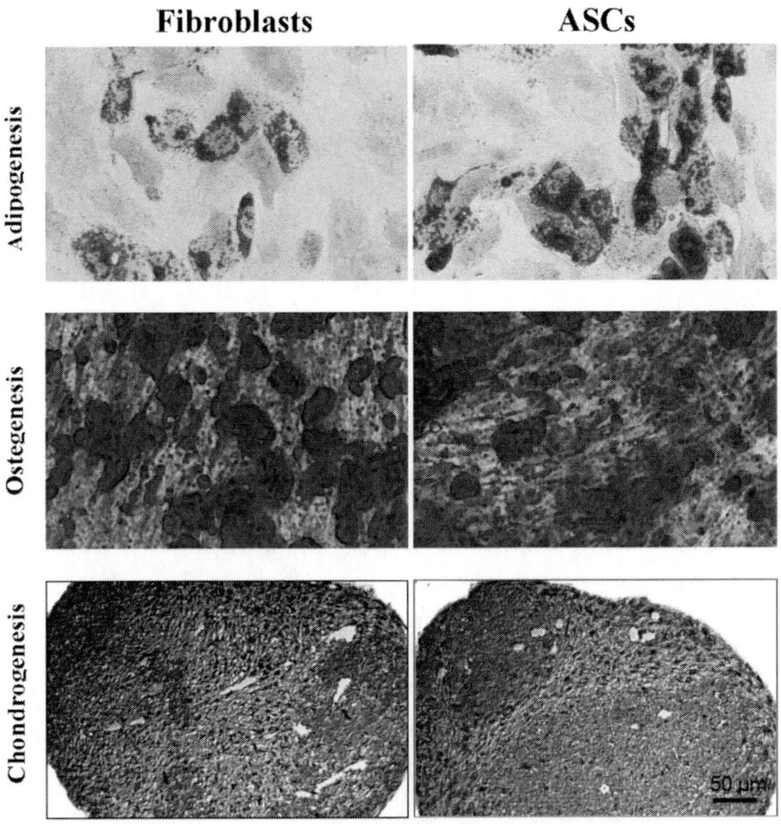

Figure 2. Adipogenesis, osteogenesis, and chondrogenesis of human dermal fibroblasts and ASCs *in vitro*. While cultured in chemically defined induction media, both fibroblasts and ASCs differentiated into adipocytes, osteoblasts, and chondrocytes. The presence of adipocytes was verified by the existence of lipid drops in cytoplasm positively stained red with Oil Red O. Osteogenesis was confirmed by calcium deposits on monolayer positively stained red by Alizarin Red S dye. Differentiated chondrocytes expressed collagen II (a specific marker of chondrocytes) with brown signals after immunohistochemistry staining (black signals represent nuclei stained by hematoxylin). ASCs showed higher adipogenic efficiency than did fibroblasts (Alt et al., 2011).

Although both fibroblasts and ASCs exhibited differentiation potential, the differences of differentiation process were also observed by several independent investigators. For instance, using RNA-seq technology, it was shown that fibroblasts and ASCs might exploit different molecular mechanisms of differentiation to reach a common cell fate. These two types of cells exhibited similar dynamics of adipogenic and osteogenic differentiation while they showed different dynamics of chondrogenic differentiation (Jaager et al., 2012). Another study showed the delayed adipogenic differentiation of fibroblasts compared with MSCs (Jaager and Neuman, 2011). The differences in the differentiation process between fibroblasts and stem cells might be attributed to their different tissue origin and/or heterogeneous populations in cell composition.

In order to further investigate the differentiation potential of fibroblasts and ASCs, single cell-derived fibroblasts and ASCs were obtained by the following clonal expansion. Cells were plated at 10 cells per cm^2 in 35-mm culture dish. Six hours after plating, single cells were marked by circling the underside of the dish with a diamond marker (4 mm diameter). When a colony was evident, the culture medium was removed and a clone ring was placed around the colony. The colony within the ring was then rinsed with PBS and subsequently detached by trypsinization. The single cell-derived cell suspension consisting of 20–50 cells was subsequently placed into a well of 24-well plate for expansion. These single cell-derived fibroblasts and ASCs were also able to differentiate into adipocytes and osteoblasts.

In contrast to the findings above, several studies showed that skin fibroblasts could not differentiate into adipocytes, osteocytes, or chondrocytes (Brendel et al., 2005; Pittenger et al., 1999). In these studies, they used foreskin fibroblasts at passage 14 for differentiation assay. The high number of passages and different sources might explain their inability to differentiate into other cell types.

Fibroblasts Secrete Angiogenic Growth Factors

Previous studies from our and other groups showed that ASCs could improve cardiac function and enhance wound repair partially via angiogenesis pathway (Altman et al.; Bai et al., 2010). Angiogenesis is the process of growth of blood capillaries from the pre-existing vascular tree. Using an *in vitro* culture system, ASCs promoted endothelial cell tube formation (Sadat et al., 2007). In addition, ASCs secreted CCL5 after stimulation with the conditioned medium from breast cancer cell line MBA231 and promoted MBA231 invasion (Pinilla et al., 2009), suggesting that the secretion capacity might also be an important characteristic for stem cells.

Fibroblast also expressed angiogenic growth factors VEGFA (vascular endothelial growth factor A), VEGFB, VEGFC, and FGF4 (fibroblast growth factors). There were no differences in the expression level of the growth factors between fibroblasts and ASCs. The secretion capacity of fibroblasts might contribute to cutaneous wound healing (Alt et al., 2011) and have potential for tissue regeneration by increasing angiogenesis.

Relationship between Fibroblasts and Stem Cells

The Origin of Fibroblasts and Stem Cells

Embryonic stem cell markers such as OCT4, KLF4, and NANOG are important for maintaining the self-renewal and pluripoteny of stem cells (Xu et al., 2009). The expression of stem cell markers and the differentiation capacity of both fibroblasts and ASCs indicate that both adipose tissue and skin contain a subset of multipotent stem cells. Accordingly, OCT4-positive stem cells were also recently identified in adult human bone marrow and hair follicles (Yu et al., 2006). These ubiquitous cells are probably deposited in peripheral tissues during embryogenesis and sustain cell turnover in their respective tissues.

Several studies have tried to identify the location of stem cells within tissues, such as adipose and skeletal muscle. However, so far there is no single marker that is specific for stem cells *in vivo*. It has been suggested that all organs and tissues might harbor ubiquitous MSCs in their perivascular niche and/or blood vessels (Baer and Geiger, 2012; Dulmovits and Herman, 2012; Klein et al., 2011; Peault, 2012).

The Relationship between Fibroblasts and Stem Cells

Li et al. investigated the relationship between MSCs and fibroblasts (Li et al., 2007). They found that BMSCs differentiated into fibroblasts and the number of bone marrow-derived myofibroblasts coincided with the development of fibrosis in a mouse doxorubicin-induced nephrosis model of chronic, progressive renal fibrosis, suggesting that MSCs are capable to differentiate into terminally differentiated fibroblasts.

Heterogeneity of Fibroblasts and Stem Cells

As we described above, over 92% fibroblasts and ASCs expressed both MSC and fibroblast markers, indicating that these tow types of cells might be homogeneous However the following experimental evidences presented in previous reports indicate that fibroblasts contain a heterogeneous cell population including stem cells with various levels of differentiation potential and terminally differentiated fibroblasts with no capacity to convert into other cell types. Firstly, the dermis is composed of various tissues including blood vessels, lymph vessels, hair follicles, and sweat glands. Secondly, the dermis might contain routinely accepted terminated differentiated cells with specific functions such as synthesizing extracellular matrix during development, wound healing, and the process of fibrosis. Thirdly, the dermis contains stem cells. Hair follicle and dermis-derived stem cells have been demonstrated the capacity to differentiate into adipocytes, osteocytes, or chondrocytes (Jahoda et al., 2003). Finally, fibroblasts are usually acquired via two different methods: (1) fibroblasts are isolated through the collagenase digestion procedure and then cultured in FBS-containing media (Johnen et al., 2006); and (2) dermis is cut into small pieces and then cultured in FBS-containing media. A few days later, spindle-like cells grow out of the explants and grow as a monolayer. These cells are considered to be fibroblasts (Villegas and

McPhaul, 2005). The isolation of ASCs is also done through a collagenase digestion procedure, as used in fibroblast preparation. Obviously, both fibroblast and stem cell preparation procedures lead to a heterogeneous cell population (including stem cells, fibroblasts, and/or pericytes). Furthermore, the data from CFU assay showed that only a low percentage of fibroblasts and ASCs could form colonies, suggesting the heterogeneous population in these cultures.

Conclusion

Dermal fibroblasts and ASCs have similar spindle-like morphology and express not only commonly accepted MSC surface markers (CD44, CD73, and CD105) (Sadat et al., 2007) but also fibroblast markers (collagen I, HSP47, vimentin, FSP, and αSMA). There is a similar gene expression pattern between fibroblasts and ASCs. In addition, both fibroblasts and ASCs are capable of forming colonies, and differentiate into tissue-specific lineages, indicating that (1) dermal fibroblasts have more plasticity than what has generally been considered; (2) fibroblasts might play an important role in cutaneous wound healing; and (3) fibroblasts might be a promising alternative autologous cell source for tissue regeneration as skin fibroblast are easily accessible. These important findings shed new light on the commonly used definition of fibroblasts and provide novel insights into many past studies performed with fibroblast preparations, since the reported effects might be greatly attributable to stem cell content within these cell preparations.

References

Alt, E., Yan, Y., Gehmert, S., Song, Y. H., Altman, A., Vykoukal, D., and Bai, X. (2011). Fibroblasts share mesenchymal phenotypes with stem cells, but lack their differentiation and colony-forming potential. Biology of the cell / under the auspices of the European *Cell Biology Organization 103*, 197-208.

Altman, A. M., Gupta, V., Rios, C. N., Alt, E. U., and Mathur, A. B. Adhesion, migration and mechanics of human adipose-tissue-derived stem cells on silk fibroin-chitosan matrix. *Acta Biomater 6*, 1388-1397.

Altman, A. M., Matthias, N., Yan, Y., Song, Y. H., Bai, X., Chiu, E. S., Slakey, D. P., and Alt, E. U. (2008a). Dermal matrix as a carrier for in vivo delivery of human adipose-derived stem cells. *Biomaterials 29*, 1431-1442.

Altman, A. M., Yan, Y., Matthias, N., Bai, X., Rios, C., Mathur, A. B., Song, Y. H., and Alt, E. U. (2008b). IFATS Series: Human adipose-derived stem cells seeded on a silk fibroin-chitosan scaffold enhance wound repair in a murine soft tissue injury model. *Stem Cells.*

Baer, P. C., and Geiger, H. (2012). Adipose-derived mesenchymal stromal/stem cells: tissue localization, characterization, and heterogeneity. *Stem cells international 2012*, 812693.

Bai, X., Sadat, S., Gehmert, S., Alt, E., and Song, Y. H. (2007). VEGF receptor Flk-1 plays an important role in c-kit expression in adipose tissue derived stem cells. *FEBS Lett 581*, 4681-4684.

Bai, X., Yan, Y., Song, Y. H., Seidensticker, M., Rabinovich, B., Metzele, R., Bankson, J. A., Vykoukal, D., and Alt, E. (2010). Both cultured and freshly isolated adipose tissue-derived stem cells enhance cardiac function after acute myocardial infarction. *Eur Heart J 31*, 489-501.

Brendel, C., Kuklick, L., Hartmann, O., Kim, T. D., Boudriot, U., Schwell, D., and Neubauer, A. (2005). Distinct gene expression profile of human mesenchymal stem cells in comparison to skin fibroblasts employing cDNA microarray analysis of 9600 genes. *Gene Expr 12*, 245-257.

Buitrago, W., and Roop, D. R. (2007). Oct-4: the almighty POUripotent regulator? *J Invest Dermatol 127*, 260-262.

Chen, F. G., Zhang, W. J., Bi, D., Liu, W., Wei, X., Chen, F. F., Zhu, L., Cui, L., and Cao, Y. (2007). Clonal analysis of nestin(-) vimentin(+) multipotent fibroblasts isolated from human dermis. *J Cell Sci 120*, 2875-2883.

Choo, A. B., Padmanabhan, J., Chin, A. C., and Oh, S. K. (2004). Expansion of pluripotent human embryonic stem cells on human feeders. *Biotechnol Bioeng 88*, 321-331.

Dulmovits, B. M., and Herman, I. M. (2012). Microvascular remodeling and wound healing: A role for pericytes. *The international journal of biochemistry & cell biology 44*, 1800-1812.

Etheridge, S. L., Spencer, G. J., Heath, D. J., and Genever, P. G. (2004). Expression profiling and functional analysis of wnt signaling mechanisms in mesenchymal stem cells. *Stem Cells 22*, 849-860.

Flavell, S. J., Hou, T. Z., Lax, S., Filer, A. D., Salmon, M., and Buckley, C. D. (2008). Fibroblasts as novel therapeutic targets in chronic inflammation. *Br J Pharmacol 153 Suppl 1*, S241-246.

Goldman, R. D., Khuon, S., Chou, Y. H., Opal, P., and Steinert, P. M. (1996). The function of intermediate filaments in cell shape and cytoskeletal integrity. *The Journal of cell biology 134*, 971-983.

Izadpanah, R., Trygg, C., Patel, B., Kriedt, C., Dufour, J., Gimble, J. M., and Bunnell, B. A. (2006). Biologic properties of mesenchymal stem cells derived from bone marrow and adipose tissue. *J Cell Biochem 99*, 1285-1297.

Jaager, K., Islam, S., Zajac, P., Linnarsson, S., and Neuman, T. (2012). RNA-Seq Analysis Reveals Different Dynamics of Differentiation of Human Dermis- and Adipose-Derived Stromal Stem Cells. *PloS one 7*, e38833.

Jaager, K., and Neuman, T. (2011). Human dermal fibroblasts exhibit delayed adipogenic differentiation compared with mesenchymal stem cells. *Stem cells and development 20*, 1327-1336.

Jahoda, C. A., Whitehouse, J., Reynolds, A. J., and Hole, N. (2003). Hair follicle dermal cells differentiate into adipogenic and osteogenic lineages. *Exp Dermatol 12*, 849-859.

Johnen, C., Hartmann, B., Steffen, I., Brautigam, K., Witascheck, T., Toman, N., Kuntscher, M. V., and Gerlach, J. C. (2006). Skin cell isolation and expansion for cell transplantation is limited in patients using tobacco, alcohol, or are exhibiting diabetes mellitus. *Burns 32*, 194-200.

Kalluri, R., and Zeisberg, M. (2006). Fibroblasts in cancer. *Nat Rev Cancer 6*, 392-401.

Kim, S. W., Cho, J. H., Hong, M. W., Rhie, J. W., and Yoon, H. R. (2008). Induction of chondrogenic differentiation in cultured fibroblasts isolated from the inferior turbinate. *Otolaryngol Head Neck Surg 139*, 143-148.

Kisiel, A. H., McDuffee, L. A., Masaoud, E., Bailey, T. R., Esparza Gonzalez, B. P., and Nino-Fong, R. (2012). Isolation, characterization, and in vitro proliferation of canine mesenchymal stem cells derived from bone marrow, adipose tissue, muscle, and periosteum. *American journal of veterinary research 73*, 1305-1317.

Klein, D., Weisshardt, P., Kleff, V., Jastrow, H., Jakob, H. G., and Ergun, S. (2011). Vascular wall-resident CD44+ multipotent stem cells give rise to pericytes and smooth muscle cells and contribute to new vessel maturation. *PloS one 6*, e20540.

Kubota, H., and Nagata, K. (2004). Roles of collagen fibers and its specific molecular chaperone: analysis using HSP47-knockout mice. *Biol Sci Space 18*, 118-119.

Li, J., Deane, J. A., Campanale, N. V., Bertram, J. F., and Ricardo, S. D. (2007). The contribution of bone marrow-derived cells to the development of renal interstitial fibrosis. *Stem Cells 25*, 697-706.

Lorenz, K., Sicker, M., Schmelzer, E., Rupf, T., Salvetter, J., Schulz-Siegmund, M., and Bader, A. (2008). Multilineage differentiation potential of human dermal skin-derived fibroblasts. *Experimental dermatology 17*, 925-932.

McAnulty, R. J. (2007). Fibroblasts and myofibroblasts: their source, function and role in disease. *Int J Biochem Cell Biol 39*, 666-671.

Midwood, K. S., Williams, L. V., and Schwarzbauer, J. E. (2004). Tissue repair and the dynamics of the extracellular matrix. *Int J Biochem Cell Biol 36*, 1031-1037.

Osonoi M, I. O., Kikuchi A, Abe S. (2011). Fibroblasts have plasticity and potential utility for cell therapy. *Hum Cell 24*, 30-34.

Park, D., Spencer, J. A., Koh, B. I., Kobayashi, T., Fujisaki, J., Clemens, T. L., Lin, C. P., Kronenberg, H. M., and Scadden, D. T. (2012). Endogenous bone marrow MSCs are dynamic, fate-restricted participants in bone maintenance and regeneration. *Cell stem cell 10*, 259-272.

Peault, B. (2012). Are mural cells guardians of stemness?: From pluri- to multipotency via vascular pericytes. *Circulation 125*, 12-13.

Pinilla, S., Alt, E., Abdul Khalek, F. J., Jotzu, C., Muehlberg, F., Beckmann, C., and Song, Y. H. (2009). Tissue resident stem cells produce CCL5 under the influence of cancer cells and thereby promote breast cancer cell invasion. *Cancer Lett 284*, 80-85.

Pittenger, M. F., Mackay, A. M., Beck, S. C., Jaiswal, R. K., Douglas, R., Mosca, J. D., Moorman, M. A., Simonetti, D. W., Craig, S., and Marshak, D. R. (1999). Multilineage potential of adult human mesenchymal stem cells. *Science 284*, 143-147.

Richardson, G. D., Arnott, E. C., Whitehouse, C. J., Lawrence, C. M., Hole, N., and Johada, C. A. (2005). Cultured cells from the adult human hair follicle dermis can be directed toward adipogenic and osteogenic differentiation. *J Invest Dermatol 124*, 1090-1091.

Sadat, S., Gehmert, S., Song, Y. H., Yen, Y., Bai, X., Gaiser, S., Klein, H., and Alt, E. (2007). The cardioprotective effect of mesenchymal stem cells is mediated by IGF-I and VEGF. *Biochem Biophys Res Commun 363*, 674-679.

Strakova, Z., Livak, M., Krezalek, M., and Ihnatovych, I. (2008). Multipotent properties of myofibroblast cells derived from human placenta. *Cell Tissue Res 332*, 479-488.

Strutz, F., Okada, H., Lo, C. W., Danoff, T., Carone, R. L., Tomaszewski, J. E., and Neilson, E. G. (1995). Identification and characterization of a fibroblast marker: FSP1. *J Cell Biol 130*, 393-405.

Villegas, J., and McPhaul, M. (2005). Establishment and culture of human skin fibroblasts. *Current protocols in molecular biology* / edited by Frederick M Ausubel et al *Chapter 28*, Unit 28 23.

Xu, B., Zhang, K., and Huang, Y. (2009). Lin28 modulates cell growth and associates with a subset of cell cycle regulator mRNAs in mouse embryonic stem cells. *Rna 15*, 357-361.

Yu, H., Fang, D., Kumar, S. M., Li, L., Nguyen, T. K., Acs, G., Herlyn, M., and Xu, X. (2006). Isolation of a novel population of multipotent adult stem cells from human hair follicles. *Am J Pathol 168*, 1879-1888.

In: Dermal Fibroblasts
Editor: Xiaowen Bai
ISBN: 978-1-62618-799-3

Chapter 9

Production of Natural Type Interferon Beta and Many Other Cytokines in Human Diploid Fibroblasts Derived from Neonatal Foreskins

*Emiko Sano**
Department of Medical Genome Sciences, Graduate School of Frontier Sciences, University of Tokyo, Japan

Abstract

A few hundred human fibroblast strains were established from neonatal foreskins and several strains were selected as good sources for natural type human interferon beta (IFN-β) production in accordance with our criteria such as growth rate, life span, karyological normality, freezing stability, and producing ability. The primary cultures of selected strains had typical fibroblasts morphorology and high IFN-β productivity. The in vitro life span was approximately 60 population doubling level (PDL) and karyotypic analysis indicated that over 90% of the cells had the normal human male diploid karyotype (46, XY). Now, one of the strains was used for producing natural human IFN-β for clinical use. Recently, it was found that the established cell lines could produce a large amount of various cytokines such as IL-6, IL-8, G-CSF, GM-CSF, GRO-α, GRO-β, IL-1α, IL-1β, IL-11, RANTES, and IFN-β in response to the stimulation of double-stranded RNA. The IFN-β and other cytokine productions were enhanced by treating the cells with a low dose of IFN-β. However, the time course of these cytokine productions was different from that of IFN-β. The production peaks of these cytokines appeared after 2 or 3 days of IFN-β production. The cytokine productions were inhibited by treating the cells with anti-human IFN-β monoclonal antibody. These data indicate that (1) various cytokine productions are mediated by previously induced IFN-β production, and (2) human fibroblasts might contribute to the host resistance by producing IFN-β and various

* Corresponding Author address: 277-8562, Kashiwanoha, Kashiwa-shi, Chiba, Japan, Email: sano_e@mgs.k.u-tokyo.ac.jp Tel: +81-4-7136-3645 Fax: +81-4-7136-3645.

physiologically active substances against virus infection, inflammation, or tumor formation.

Keywords: Human diploid fibroblasts, IFN-β, cytokine, double-stranded RNA, PDL, diploidy

Introduction

Fibroblasts are the most common cells of connective tissue in animals (Horoszewicz et al., 1978). It has been reported that fibroblasts could produce interferon beta in response to viral infection or the stimulation of synthetic double-stranded RNA using the cells derived from mouse or human tissue (Sano et al., 1974; Van Damme et al., 1981). Thereafter, it has been demonstrated that fibroblasts could also produce varying cytokines and hematopoietic growth factors such as interleukin-6 (IL-6) (Kohase et al., 1987), IL-8 (Van Damme et al., 1989) and colony stimulating factors (CSF) such as G-CSF, M-CSF, and GM-CSF (Fibbe et al., 1988) in response to the stimulation of synthetic double-stranded RNA, polyriboinosinic-polyribocytidylic acid (poly I : poly C). Double-stranded RNA is produced by most viruses at some point during their replication and is associated with viral infection (Jacobs and Langland, 1996).

IFNs are a family of related cytokines that mediate a wide range of diverse functions including anti-viral, anti-proliferative, anti-tumor, and immuno-modulatory activities. IFNs are currently classified into two major groups of type I and type II (Gresser, 1990). IFN-α, -β, and –ω are included in type I IFN. The type II IFN is designated IFN-γ, also known as immune IFN. These two types of IFNs bind to distinct cellular receptors and activate both individual and overlapping pathways. Recently, some molecules relating to the activation of IFN-β gene and subsequent IFN-β inducible genes (e.g., cytokine and chomkine genes) induced by double-stranded RNA have been identified. Toll-like receptor 3 (TLR3) was identified as a receptor for double-stranded RNA produced by viruses and it was certified to be necessary for IFN-β gene activation (Alexopoulou et al., 2001).

Following IFN-β gene activation, some cytokine mRNAs such as RANTES, MCAF/MCP-1, and IP-10 were also demonstrated to be activated (Yamamoto, et al., 2002). Thereafter, some molecules relating down-stream pathway of TLR3 have been identified (Yamamoto et al., 2003; Oshiumi et al., 2003). However, very little is known about the production mechanisms of many other biological substances detected in fibroblasts.

IFN-β has been produced by a super-induction method. The method has been devised as an effective production method for IFN-β by treating the cells with the metabolic inhibitores such as cycloheximide and actinimycin-D after poly I : poly C stimulation (Billiau et al., 1973; Vilcek et al., 1976). In this method, IFN-β production was enhanced by inhibiting the synthesis of represser proteins appeared later than IFN-β using metabolic inhibitors.

It has been demonstrated that the fibroblasts derived from neonatal foreskins could produce high level of human IFN-β (Lee et al., 2006); therefore, foreskin-derived fibroblasts appear to be suitable cells for preparation of human IFN-β for clinical application.

However, the large scale culture of the cells has been difficult for their anchorage dependent growth. In our lab, we have established several human neonatal foreskin fibroblast strains for IFN-β production and could construct a microcarrier culture system for large scale cultures of these fibroblasts (Sano et al.,1987).

Recently, we found that foreskin derived fibroblasts can also produce various kinds of cytokines (e.g., IL-1α, IL-1β, IL-6, IL-8, IL-11, G-CSF, GM-CSF), followed by the stimulation of poly I : poly C. These results indicate that fibroblast cells may contribute to the self-defense mechanisms in a body by producing these biological substances.

In this report, we introduced our works regarding the establishment of functional human diploid fibroblast strains derived from neonatal foreskins and productions of IFN-β and many other substances. The production mechanisms of cytokines will also be demonstrated.

Methods

This work was carried out in the basic research laboratories of Toray Industries, Kanagawa, Japan, and Proteios Research Inc. Kanagawa, Japan.

Establishment of Cell Strains

A few hundred fibroblast strains were established from neonatal foreskins kindly supplied from Dr. R.E. Gilis of the Providence Hospital, Anchorage, Alaska. Several strains were selected from these strains as good sources for human IFN-β production in accordance with our criteria (Table 1). The cells of 6 population doubling level (PDL), 14 PDL, and 22 PDL were stored at the concentration of 2×10^6 cells/ampoule in a liquid nitrogen *container* and a freezer at 80°C, separately.

Table 1. Criteria for selection of functional human diploid fibroblast strains

		Ranking				
		A	B	C	D	E
Doubling time	Hr	24-48	48-72	72-96	96-120	>120
Density	10^4/cm^2	>7	3-7	0.7-3	0.07-0.7	<0.07
Split ratio (a)	Fold	>32	32-16	16-8	8-4	4-2
Life span	PDL	>50	50-40	40-30	30-20	<20
Serum requirement	%	<5	5-10	10-15	15-20	>20
Viability(liquid N_2) (b)	%	>90	90-80	80-70	70-50	<50
Diploidy	%	>90	90-80	80-70	70-60	<60
Stability (c)	Days	>14	14-10	10-6	6-3	<3
IFN-β production	IU/mL	>5,000	5-3,000	3-1,000	1,000-200	<200

(a)Dilution ratio able to form complete confluent culture.
(b)Viability of 1×10^6 cells stored in liquid N_2 after a week.
(c)Maintenance days of confluent culture grown in culture flasks.

Cell Cultures

The cells were maintained in culture flasks (Corning, NY) routinely and experiments were usually performed using plastic tissue culture flasks and 6-well plates (Iwaki Glass,

Tokyo, Japan). An efficient microcarrier culture system established in our laboratory (Sano et al., 1987; Sano et al., 1988; Ohashi et al., 2003) was also employed as a stable production system for IFN-β and other cytokine productions at 200 mL scale. The growth medium for both the monolayer and microcarrier cultures was Eagle's minimum essential medium (MEM, Nissui Pharmaceutical, Tokyo, Japan) supplemented with 5% fetal calf serum (FCS, Life Technologies, Grand Island, NY). Experiments were usually performed using the cells of 20-45 PDL with diploidy of over 95%.

Production of IFN-β and Other Products

Confluent cells were primed for 24 hr at 37°C with 100 international units (IU)/mL of IFN-β (Toray Industries, Tokyo, Japan) in Eagle's MEM. The cells were treated with 10 μg/mL of poly I : poly C (Yamasa Corporation, Chiba, Japan) for 1hr and incubated for 5 to 6 days for the release of IFN-β and other products. For the comparison of productivity, a super-induction method was also used. The super-induction was carried out according to the procedure described by Vilcek et al. (1976). The cells primed with 100 IU/mL of IFN-β were stimulated with 10 μg/mL of poly I : poly C and treated with 5 μg/mL of cycloheximide (Wako Pure Chemical, Osaka, Japan) for 4 hr and with 4μg/mL of actinomycin-D (Makor Chemical, Jerusalem, Israel) for 1 hr.

Quantifications of IFN-β and Other Products

The quantities of IFN-β and other products were determined by ELISA systems. The kits for IFN-β, IL-6, and IL-8 were developed in Toray Industries, Tokyo, Japan, (Yamazaki et al., 1989; Ida et al., 1990; Ida et al., 1992) and the yields of GM-CSF, G-CSF, IL-1α, IL-1β, IL-11, RANTES, and TNF-α were assayed by the ELISA kits purchased from R&D Systems (Minneapolis, MN). The amounts of GRO-α and GRO-β were determined by the kits developed by Immuno-Biological Laboratories (Gunma, Japan).

Northern Blot Analysis

Poly(A)$^+$RNA was isolated from the disruption of fibroblast cells (FC1) stimulated with poly I : poly C after priming the cells with IFN-β by incubating with oligo-(dT) cellulose, and washing and eluting retained poly(A)$^+$RNA using Fast Track mRNA isolation kit (Invitrogen, San Diego, CA). Poly(A)$^+$RNA was separated by gel electrophoresis and transferred overnight to nylon filters (Hybond N, Amersham Pharmacia Biotech, Piscataway, NJ). Northern blot hybridization was carried out using nucleotide fragments labeled with [α-^{32}P] dCTP as probes, and hybridized filters were autoradiographed. Probes used were full-length IL-6, GM-CSF, and IFN-β cDNA. The filter was analyzed using Bioimage Analyzer, BAS 2000 (Fuji Photo Film Co., Tokyo, Japan).

Neutralization of IFN-β Activities by Monoclonal Antibody

A mouse anti-human IFN-β1 monoclonal antibody used to neutralize poly I : poly C-induced IFN-β production was prepared in Toray Industries (Tokyo, Japan). A mouse anti-human MCAF monoclonal antibody prepared by Toray Industries (Tokyo, Japan) was used as a control antibody of the same iso-type as anti-IFN-β. The cells primed with IFN-β were treated with poly I : poly C in the presence of various concentrations of anti-human IFN-β antibody.

Chromosome Number and Karyotype Analysis

The chromosome number and karyotypic analysis were carried out according to the slightly modified procedures reported by Imai (1975) and Yoshida et al. (1984). The growing cells were treated with the hypotonic salt solution and fixed on the clean slides. The slides were used for chromosome number and karyotypic analysis by lining up and counting the chromosomes of homologous pair observed by G-band staining using Giemsa solution (Sigma-Aldrich, St.Louis, MO) according to Seabright's method (Seabright, 1971) with a slight modification by Yoshida et al. (1984).

Tumorigenicity Test

12-week old male mice (nu/nu) were used. 1×10^6 of Hela cells and 1×10^7 of FC-1 cells were inoculated subcutaneously into each of 5 nude mice. The number of mice formed tumors with the size of about $2 \times 1 \times 0.5$ cm^3 was counted after 24 days of cell inoculation.

Results

Selection and Characterization of Cell Strains for IFN-β Production

A few hundred fibroblast strains have been established from human neonatal foreskins. Several strains were selected as good producing cells according to the criteria shown in Table 1. Five rankings were set up to evaluate the response of the cell strains; however, it has been difficult to find cell strains which achieve a high ranking for all items. Several strains demonstrated "A" abilities in most items and "B" abilities in a few. These cell strains were selected as candidates for the production of human IFN-β. Ampoule stocks of the candidate cells at 6 PDL were kept in liquid nitrogen tank as the primary cell stocks. The characteristics of the three promising cell strains (FC-1, FC-2, and FC-3) were shown in Table 2. These cells had typical fibroblast morphology and grew well to form a uniform monolayer. The in vitro life span was approximately 60 PDL. As shown in Fig. 1, karyotypic analysis indicated that over 90% of the cells (even in older generations) had the normal diploid chromosomes for the human male karyotype (46, XY). Furthermore, the obvious chromosomal changes such as the deletion and the translocation were not detected in these cells. Table 3 shows that FC-1 (32

PDL) exhibited no tumorigenicity in nude mice. When 1×10^6 HeLa cells were inoculated into each of 5 nude mice by subcutaneous injection, tumor formation was clearly observed in all mice tested. Whereas, even the injection of 1×10^7 FC-1 cells per mouse did not cause tumor formation. These cells were free of detectable adventitious agents such as viruses, mycoplasma, bacteria, and fungi. Therefore, these cell strains were expected as the primary sources for large-scale preparation of human IFN-β for clinical use.

Priming Effects of IFN-β on IFN-β and other Cytokine Productions

IFN-β and other cytokines were produced by stimulating the cells with poly I : poly C after priming with a low dose of IFN-β (100 IU/mL) as described in METHODS. The priming effect of IFN-β on the production of IFN-β and other cytokines was examined using FC-1 cells. As shown in Fig. 2, the production of IFN-β and other products increased 5- to -40 fold by treating the cells with 100 IU/mL of IFN-β for 18-24 hr. The priming effect by type 1 IFN has been well known in IFN-β production by poly I : poly C. However, the effects on cytokine productions were found for the first time in this study. These results indicated the existence of the common synthesis mechanism in IFN-β and cytokine productions.

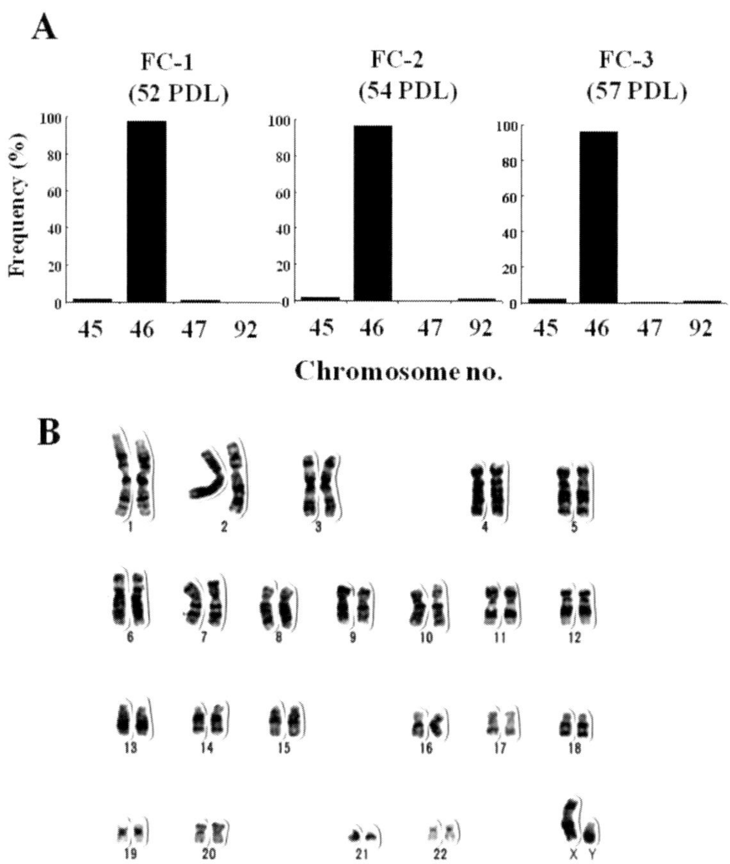

Figure 1. Karyotypic analysis. A: Distribution of chromosome numbers of aged FC-1, FC-2, and FC-3 cells. B: The homologous pairs classified by G-band staining.

Table 2. Ranking of FC cells in each item for selection

	Ranking		
	FC-1	FC-2	FC-3
Doubling time	A	A	A
Density	A	A	A
Split ratio	A	B	B
Life span	A	A	A
Serum requirement	A	A	A
Viability in liquid N_2	B	B	B
Diploidy (Karyology)	A	A	A
Stability	B	A	B
IFN-β productivity	A	A	A

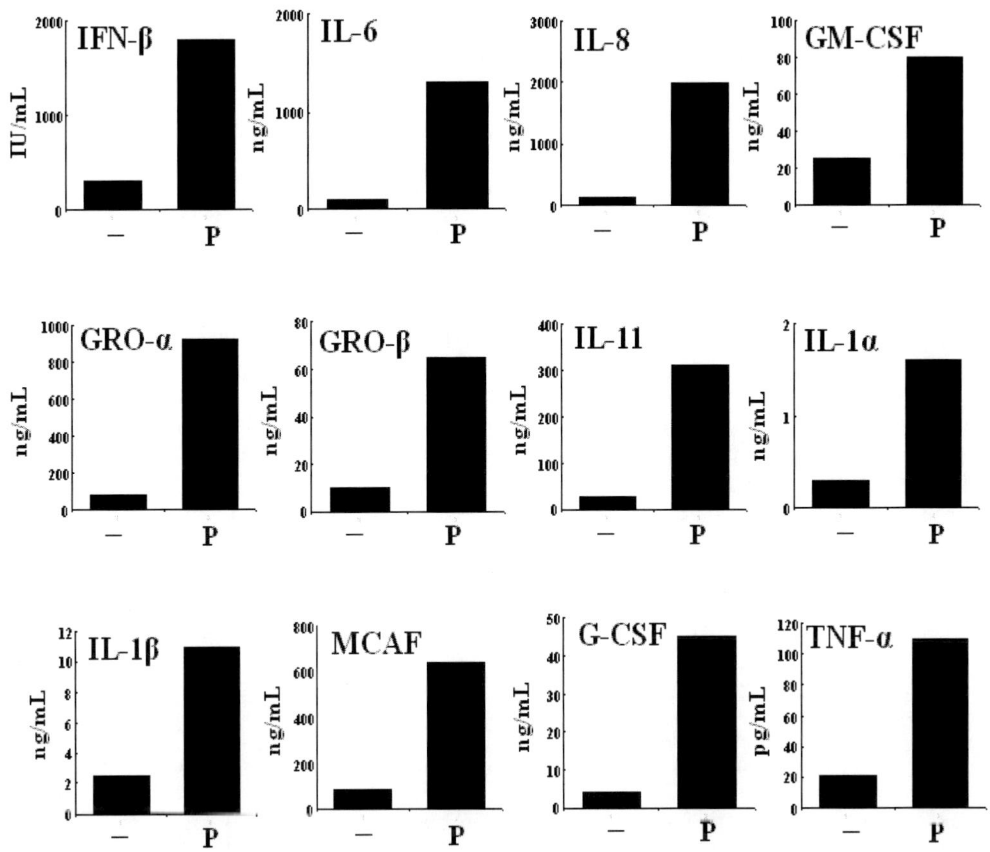

Figure 2. Priming effects of IFN-β on IFN-β and other cytokine productions. The confluent FC-1 cells grown in microcarrier cultures were treated with 100 IU/mL of IFN-β for 24 hr at 37°C prior to poly I : poly C stimulation (10 μg/ml). Culture supernatants were harvested at 2 days for IFN-β and at 5 days for other cytokines after stimulation. The contents of IFN-β and other cytokines were determined by ELISA kits.(-), not priming; (P), priming. (Referred from J Cell Biochem (2007) 100:1464).

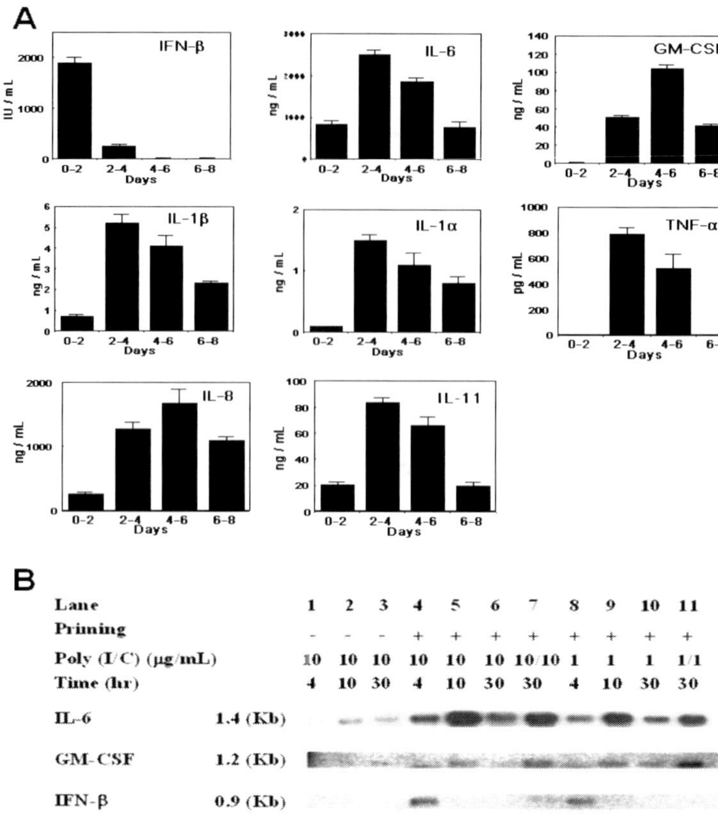

Figure 3. Production time courses of IFN-β and other cytokines. A: IFN-β and other cytokine productions. The supernatants in microcarrier cultures were harvested every 2 days for the determination of cytokine contents. B: Northern blot analysis. The expression patterns of IFN-β, IL-6, and GM-CSF mRNA under various conditions were examined using FC-1 cells. (Referred from J Cell Biochem (2007) 100:1465).

Table 3. Tumorigenicity of FC-1 cells in nude mice

Cells	Cells/Mouse[(a)]	Tumor mice/Tested mice
Hela	10^6	5/5[(b)]
FC-1 (32 PDL)	10^7	0/5[(c)]

(a) Mice used are 12-week-old males (nu/nu).
(b) At 24 days after SC inoculation of the cells, the tumors with the size of about 2 × 1 × 0.5 cm^3 occurred in all mice tested.
(c) Observation was carried out for 50 days after the inoculation.

Differential Production Profile of IFN-β and other Cytokines

Subsequently, the production time courses were examined. As shown in Fig. 3A, the maximum production of IFN-β was observed at 24-48 hr after poly I : poly C stimulation; however, the production peaks of other cytokines appeared at 2 or 3 days which were later than IFN-β production. Large amounts of cytokines were produced in this study while IFN-β yield is little. The results were confirmed by Northern blot analysis for the expression of IFN-β, IL-6, and GM-CSF mRNA. As shown in Fig. 3B, the expression peaks of IL-6 and GM-CSF mRNA appeared at 10 hr after poly I : poly C stimulation was later than that of IFN-β (4 hr). While, most cytokines were produced very little in super-induction method contrary to IFN-β production as previously reported (Emiko Sano et al., 2007). These results indicate that cytokine production mechanisms in fibroblasts stimulated with double-stranded RNA are different from those of IFN-β.

Cytokine Productions Mediated by IFN-β

The results of Fig. 3A and B indicate the autogenous IFN-β might be involved in the production processes of other cytokines although the priming effects by IFN-β are common. To certify this concept, the effects of neutralizing inducible IFN-β on the various cytokine productions were examined using specific monoclonal antibody against human IFN-β. FC-1 cells primed with 100 IU/mL of IFN-β were stimulated with 10 μg/mL of poly I : poly C in the presence of various concentrations of anti-human IFN-β antibody. As shown in Fig. 4, the productions of IL-6, IL-8, GM-CSF, GRO-β, RANTES, and IFN-β were inhibited by the addition of IFN-β antibody in a dose dependent manner. The effects on the productions of IL-1α, IL-1β, TNF-α, IL-11, G-CSF, GRO-α, and MCAF were also suppressed by the anti-IFN-β monoclonal antibody (data not shown). Whereas, the cytokine productions by poly I : poly C was not enhanced by the external addition of IFN-β, indicating that only autogenously induced IFN-β can induce various biological substances. In conclusion, IFN-β produced in dermal fibroblasts may play a key role in cytokine productions and contribute to self-defense against various diseases.

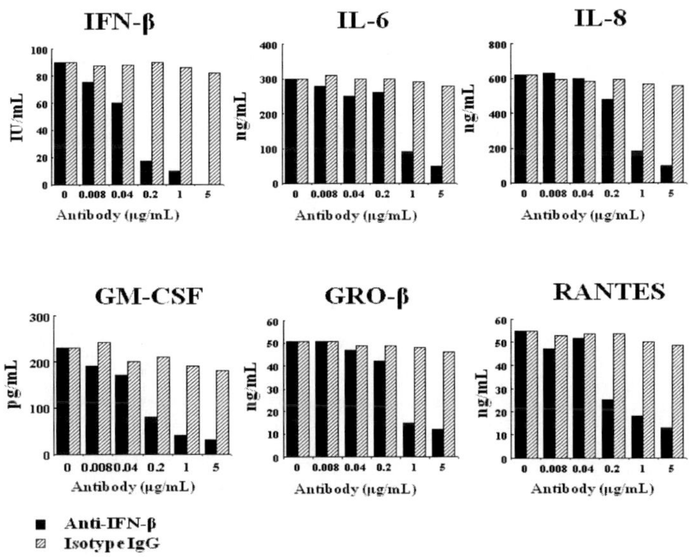

Figure 4. Inhibition of IFN-β and other cytokine productions by anti-human IFN-β antibody. FC-1 cells grown in 6-well plates were primed with IFN-β (100 IU/mL) for 24 hr and stimulated with poly I : poly C (10μg/mL) in the presence of various concentrations of anti-IFN-β antibody. The corresponding dosage of MCAF monoclonal antibody was used as iso-type IgG control. (Referred from J Cell Biochem (2007) 100:1468.

References

Alexopoulou, L., Holt, A. C., Medzhitov, R., Flavell, R. A. (2001). Recognition of double-stranded RNA and activation of NF-kB by Toll-like receptor 3. *Nature 413*, 732-738.

Billiau, A., Joniau, M., and De Somer, P. (1973). Mass production of human interferon in diploid cells stimulated by poly I : poly C. *Journal of General Virology 19*, 1-8.

Fibbe, W. E., Van Damme, J., Billiau, A., Duinkerken, N., Lurvink, E., Ralph, P., Altrock, B, W., Kaushansky, K., Willemze, R., and Falkenburg, J. H. F. (1988). Human fibroblasts produce Granulocyte-CSF, Macrophage-CSF, and Granulocyte-Macrophage- CSF following stimulation by Interleukin-1 and Poly (rI). Poly (rC). *Blood. 72*, 860-866.

Gresser, I. (1990). Biologic effects of interferons. *Journal of Investigative Dermatology. 95*, 66s-71s.

Horoszewicz, J. S., Leong, S. S., Ito, M., Di Berardino, L., Carter, W.A. (1978). Aging in vitro and large-scale interferon production by 15 new strains of human diploid fibroblasts. *Infection and Immunity 19*, 720-726.

Ida, N., Sakurai, S., Hosaka, T., Kunitomo, T., Matsuura, T., Kohase, M. (1990). An enzyme-linked immunosorbent assay for the measurement of human interleukin-6. *Journal of Immunological Methods 133*,279-284.

Ida, N., Sakurai, S., Hosoi, K., Kunitomo, T. (1992). A highly sensitive enzyme-linked immunosorbent assay for the measurement of interleukin-8 in biological fluid. *Journal of Immunological Methods 156*, 27-38.

Imai, H. T. (1975). Evidence for non-random localization of the centromere on mammalian chromosomes. *Journal of Theoretical Biology 49*, 111-123.

Jacobs, B. L., and Langland, J. O. (1996). When two strans are better than one: the mediators and modulators of the cellular responses to double-strended RNA. *Virology 15;219*, 339-349.

Kohase, M., May, L., Tamm., E., Vilcek, J., and Sehgal, P. B. (1987). A cytokine network in human diploid fibroblasts: Interactions of β-interferon, tumor necrosis factor, platelet-derived growth factor, and interleukin-1. *Molecular Cellular Biology 7*, 273-280.

Lee, G. C., Yi, H. A., and Lee, C. H. 2006. Stimulation of interferon-beta gene expression by human cytomegalovirus via nuclear factor kappa B and phosphatidylinositol 3-kinase pathway. *Virus Research 117*:209-214.

Ohashi, K., Naruto, M., Nakaki, T., Sano, E. (2003). Identification of interleukin-8 converting enzyme as cathepsin L. *Biochimica Et Biophysica ACTA 1649*, 30-39.

Oshiumi, H., Matsumoto, M., Funami, K., Akazawa, T., Seya, T. (2003). TICAM-1, an adapter molecule that participates in Toll-like receptor 3-mediated interferon-β induction. *Nature Immunology 4*, 161-167.

Sano, E., Matsui, Y., and Kobayashi, S. (1974). Production of mouse interferon with high titers in a large-scale suspension culture system. *Japanese Journal of Microbiology 18*, 165-172.

Sano, E., Iizuka, M., and Kobayashi, S. (1987). Effect of glucocorticoid hormones on growth of human fibroblast cells and interferon production in a microcarrier culture system. *Cell Structure and Function 12*, 509-517.

Sano, E., Okano, K., Sawada, R., Naruto, M., Sudo, T., Kamata, K., Iizuka, M., and Kobayashi, S. (1988). Constitutive long-term production and characterization of recombinant human interferon-gamma from two different mammalian cells. *Cell Structure and Function 13*, 143-159.

Sano, E., Ohashi, K., Sato, Y., Kashiwagi, M., Joguchi, A., and Naruse, N. (2007). A possible role of autogenous IFN-β for cytokine productions in human fibroblasts. *Journal of Cellular Biochemistry 100*, 1459-1476.

Seabright, M. (1971). A rapid banding technique for human chromosomes. *Lancet* 2, 971-972.

Van Damme, J., and Billiau, A. (1981). Large-scale production of human fibroblast interferon. *Methods in Enzymology 78*, 101-119. .

Van Damme, J., Decock, B., Conings, R., Lenaerts, J., P., Opdenakker, G., and Billiau, A. (1989). The chemotactic activity for granulocytes produced by virally infected fibroblasts is identical to monocyte-derived interleukin 8. *European Journal of Immunology 19*, 1189-1194.

Vilcek, J., Havell, E. A., and Kohase, M. (1976). Super-induction of interferon with metabolic inhibitors: Possible mechanisms and practical applications. *Journal of Infectious Diseases* (Suppl) *133,* A22-A29.

Yamamoto, M., Sato, S., Mori, K., Hoshino, K., Takeuchi, O., Takeda, K., Akira, S. (2002). Cutting edge: A noble Toll/Il-1 receptor domain-containing adapter that preferentially activates the IFN-beta promoter in the Toll-like receptor signaling. *Journal of Immunology* 169, 6668-6672.

Yamamoto, M., Sato, S., Hemmi, H., Hoshino, K., Kaisho, T., Sanjo, H., Takeuchi, O., Sugiyama, M., Okabe, M., Takeda, K., Akira, S. (2003). Role of adapter TRIF in the MyD88-independent Toll-like receptor signaling pathway. *Science 301*, 640-643.

Yamazaki, S., Kuramoto, K., Matsuo, E., Hosoi, K., Hara, M., and Shimizu, H. (1989). Highly sensitive enzyme immunoassay. *Journal of Immunoassay 10*, 57-73.

Yoshida, T. H., Sadaie, T., and Sadaie, Y. (1984). Somatic and meiotic chromosomes of the small free-living nematode, *Caenorhabditis elegans. Proceedings of the Japan Academy*, Series B, Physical and Biological Sciences *60*, 54-55.

In: Dermal Fibroblasts
Editor: Xiaowen Bai

ISBN: 978-1-62618-799-3

Chapter 10

Reprogramming of Dermal Fibroblasts to Induced Pluripotent Stem Cells and Cardiomyocytes

***Dongrui Ma*[1] *and Reginald Liew*[1, 2 *]**
[1]Research and Development Unit,
National Heart Center of Singapore, Singapore
[2]Duke-NUS Graduate Medical School, Singapore

Abstract

Reprogramming of induced pluripotent stem (iPS) cells has been successfully achieved from mouse and human somatic cells. iPS cells have the capability of indefinite self-renewal and can be differentiated into almost all cell types of the body. This makes them a valuable tool for studying developmental biology, modeling of human genetic diseases, and investigating their therapeutic potential. Specifically, human iPS cells were reprogrammed from dermal fibroblasts from patients with heart diseases, and used to generate patient-specific cardiomyocytes, which have termendous applications in biomedical research, drug development, and clinical practice.

Keywords: Dermal fibroblasts, induced pluripotent stem cells, cardiac differentiation

Introduction

The recent breakthrough of generating induced pluripotent stem (iPS) cells from human skin fibroblasts, has opened a new era of opportunities for biomedical research and clinical applications (Takahashi et al., 2007; Yu et al., 2007). The iPS cells have extensive expansion capacity and the potential to differentiate into all cell types from all three primary germ

[*] Corresponding Author Address: Duke-NUS Graduate Medical School Singapore, 8 College Road, Singapore 169857; Email: reginald.liew@duke-nus.edu.sg.

layers, including cardiomyocytes (CMs). The generation of iPS cells and derivation of CMs from these cells have multiple applications. Firstly, access to an *in vitro* model of human heart development allows investigators to study human heart development processes and mechanisms which were not previously possible. Secondly, the ability to generate iPS cells and CMs from patients with inherited cardiac diseases provides an unparalleled opportunity to study these diseases *in vitro*; and makes it possible for the pharmaceutical industry to screen and test drugs for cardiovascular diseases using living, functional human CMs (Inoue and Yamanaka, 2011). In a few recent studies, iPS cells have been reprogrammed from patients with Long QT syndrome type I and type II, and arrythymogenic right ventricular cardiomyopathy and CMs have been derived from these patient-specifc iPS cells (Itzhaki et al., 2011; Ma et al., 2012; Moretti et al., 2010; Yazawa et al., 2011). These CMs reserved the cellular phenotype of the patients' heart diseases. Hence these CMs are suitable tools to be used for testing the effects of drugs and other therapeutic interventions on these diseases *in vitro*. Finally, the generation of clinical grade iPS cells and related CMs provides a powerful approach to repair or replace the injured heart tissue(Oh et al., 2012). In this chapter, we describe the current methods for the generation of iPS cells from human dermal fibroblasts using retroviral transfection and non-viral methods. In addition, we discuss the techniques used to induce iPS cells to differentiate into functional CMs.

Current Methods of Reprogramming Induced Pluripotent Stem Cells

One of the most commonly used methods of reprogramming somatic cells to iPS cells utilizes viruses (retroviruses or lentiviruses) to deliver reprogramming factors into the host cells (e.g., dermal fibroblasts). Over the past few years, alternative reprogramming methods have been developed to generate iPS cells (Figure 1). These alternative methods are aimed at establishing safer iPS cells with minimal genomic alterations and improving reprogramming kinetics and efficiency. iPS cells have been generated using integrating and excisable lentiviral and transponson systems (Chang et al., 2009; Somers et al., 2010; Woltjen et al., 2009) or non-integrating adenoviral, episomal and minicircle plasmid systems (Okita et al., 2008; Si-Tayeb et al., 2010; Stadtfeld et al., 2008; Yu et al., 2009; Zhou and Freed, 2009). iPS cells have also been generated via non-integrating and DNA-free methods: Sendai virus (Fusaki et al., 2009); recombinant proteins (Kim et al., 2009; Zhou et al., 2009), micro-RNAs (Anokye-Danso et al., 2011; Miyoshi et al., 2011), and synthetic modified messenger RNA (Warren et al., 2010).

Advantages and Disadvantages of the Different Methods of iPS Cell Generation

The viruses used for reprograming integrate randomly into the genome of the host cells, which can potentially result in insertional mutagenesis. Thus, an important disadvantage of this technique is the restriction of the differentiation capacity of the reprogrammed cells (Okita and Yamanaka, 2010) and an increased risk of tumorigenicity (Okita et al., 2007).

Alternative protocols of reprogramming iPS cells with virus-free and transgene-free methods also have considerable limitations. For example, excisable lentiviral and transposon systems may cause genomic disruptions during integration and leave residual transgenes after excision. Adenoviral and plasmid systems utilize DNA-based constructs which still pose some risk of genomic recombination or insertional mutagenesis. Thus, systematic characterization is required to rule out any random integration of the constructs. Sendai virus has an RNA genome which makes it difficult to purge reprogrammed cells of replicating virus. Recombinant proteins have short half-lives and the protocol involved has a technical challenge as it requires the generation of large quantities of pure proteins and the multiple administrations of proteins. Furthermore, many of these alternative reprogramming methods have lower reprogramming efficiencies than the original viral reprogramming method. Inefficiency of reprogramming increases the cost and difficulty of deriving iPS cells, posing a problem when working with rare cell sources.

Here, we describe in detail the current methods we have employed in our lab for the generation of iPS cells using viruses. Figure 2 shows a schematic diagram of the stages involved.

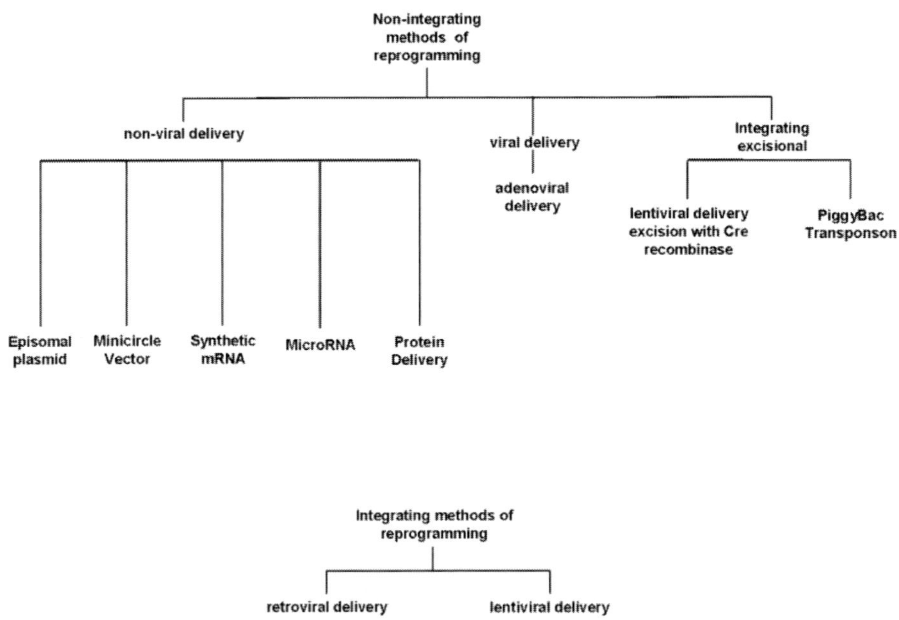

Figure 1. Summary of the different methods currently available for reprogramming somatic cells into iPS cells. These methods are classified based on whether or not the reprogramming factors are integrated into the host genome.

Culture of Human Skin Dermal Fibroblasts

We obtain a skin punch biopsy from the patient following informed consent. The skin dermis is then manually separated and minced. Dermal pieces are next plated onto a petri-dish with Dulbecco's modified eagle medium (DMEM, Invitrogen, USA) supplemented with 10% fetal bovine serum (FBS, Invitrogen, USA), 0.5% penicillin and streptomycin

(Invitrogen, USA). Fibroblasts migrate out of the dermis within 7 days. After fibroblasts reach 80-90% confluence, fibroblasts are subcultured and maintained weekly.

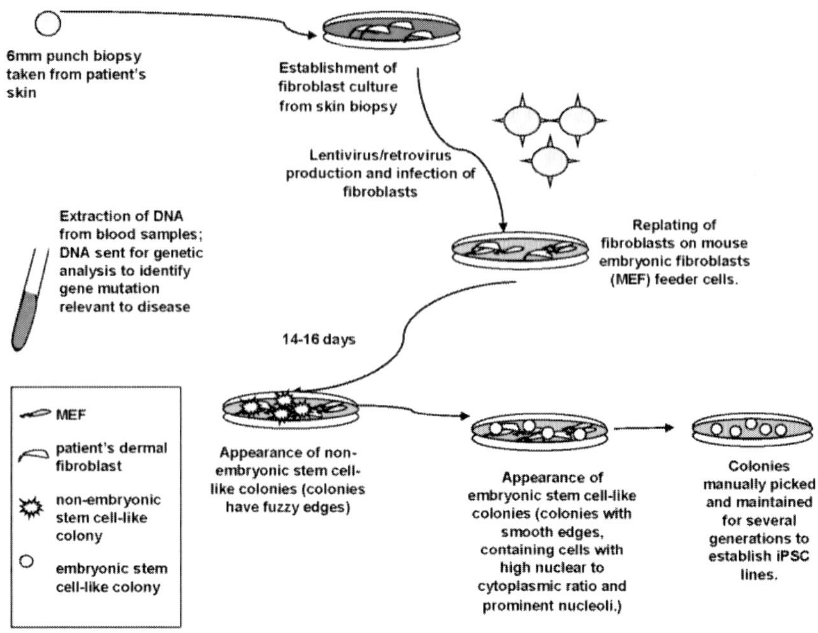

Figure 2. Schematic diagram showing the generation of iPS cells from patient's dermal fibroblasts. Non-embryonic stem cell-like colonies which appear first are not picked. Only embryonic stem cell-like colonies which appear later are picked and maintained in culture.

Reprogramming of iPS Cells

HEK 293GP cells (Cell Biolabs, USA) are maintained in DMEM containing 10% FBS, 2mM L-glutamine (Invitrogen, USA), 1 × 10^{-4}M nonessential amino acids (Invitrogen, USA), 0.1 mM 2-mercaptoethanoel (Invitrogen, USA), 1 mM sodium pyruvate (Sigma- aldrich, USA), and 0.5% penicillin and streptomycin. A calcium phosphate transfection protocol is used to package the retrovirus in 293GP cells. pMXs-Sox2, pMXs-Oct3/4, pMXs-Klf4 and pMXs-C-Myc plasmids are purchased from Addgene. 16 hours after transfection, the cell culture medium is replaced with fresh medium. 24 hours later, the medium is collected as the first virus-containing supernatant, and fresh medium is supplemented. 48 hours later, the supernatant is collected as the second virus- containing medium. Human fibroblasts are seeded into 6-well plate at a density of 1×10^5 cells/well 1 day before virus infection. The virus-containing supernatants are supplemented with 8μg/ml polybrene (Sigma aldrich, USA). Equal amounts of supernatants containing each of the four retroviruses are mixed, transferred to the fibroblast dish, and incubated overnight. 24 hours after infection, the virus-containing medium is replaced with the second virus supernatant. Six days after transduction, fibroblasts are harvested by trypsinization and replated onto a Mouse embryonic feeder (MEF) layer. The next day, the medium is replaced with embryonic stem (ES) cell culture medium (DMEM/F12 supplemented with 20% knockout serum replacement (Invitrogen,

USA), 2mM L-glutamine (Invitrogen, USA), 1 × 10^{-4}M nonessential amino acids, 0.1 mM 2-mercaptoethanoel (Invitrogen, USA), 10ng/ml beta-fibroblast growth factor (βFGF) and 0.5% penicillin and streptomycin. The cell culture medium is changed every other day. 30 days after transduction, ES cell like colonies are manually picked up and transferred into fresh MEF feeder in ES cell medium for expansion and downstream studies.

Characterization if Induced Pluripotent Cells

The following experiments are performed to confirm the pluripotency of the reprogrammed cells from dermal fibroblasts: (1) pluripotent stem cell-marker (e.g., OCT-4) expression using immunofluorescence staining and PCR and (2) teratoma formation in SCID mice to induce teratoma in SCID mice which consist of all three germ layers.

Table 1. Primer Sequences

Primer	Sequence (5' to 3')
Oct4	Forward: TAT GCA CAA CGA GAG GAT TTT G
	Reverse: TGT GTC CCA GGC TTC TTT ATT T
Sox2	Forward: CAC CTA CAG CAT GTC CTA CTC G
	Reverse: GGT TTT CTC CAT GCT GTT TCT T
Klf4	Forward: ACA AAG AGT TCC CAT CTC AAG G
	Reverse: TGG ATT CCT CAT TTT TCC TGA T
c-Myc	Forward: CTG AGA CAG ATC AGC AAC AAC C
	Reverse: TCT TTT ATG CCC AAA GTC CAA T
Lin28	Forward: AAA AGG AAA GAG CAT GCA GAA G
	Reverse: GTT GGA GTT CCT TTG ACT CAC C
Nanog	Forward: GAC TGA GCT GGT TGC CTC AT
	Reverse: TTT CTT CAG GCC CAC AAA TC
DPPA2	Forward: CCG TCC CCG CAA TCT CCT TCC ATC
	Reverse: ATG ATG CCA ACA TGG CTC CCG GTG
DPPA4	Forward: GGA GCC GCC TGC CCT GGA AAA TTC
	Reverse: TTT TTC CTG ATA TTC TAT TCC CAT
ESG1	Forward: ATA TCC CGC CGT GGG TGA AAG TTC
	Reverse: ACT CAG CCA TGG ACT GGA GCA TCC
GDF3	Forward: CTT ATG CTA CGT AAA GGA GCT GGG
	Reverse: GTG CCA ACC CAG GTC CCG GAA GTT
NAT1	Forward: AGG GTT CGG GGG AGG TAA GGG TGC
	Reverse: AGG GTT GCG TGC GTA AAG CCG GAG
NODAL	Forward: GGG CAA GAG GCA CCG TCG ACA TCA
	Reverse: GGG ACT CGG TGG GGC TGG TAA CGT TTC
REX1	Forward: CAG ATC CTA AAC AGC TCG CAG AAT
	Reverse: GCG TAC GCA AAT TAA AGT CCA GA
UTF1	Forward: CCG TCG CTG AAC ACC GCC CTG CTG
	Reverse: CGC GCT GCC CAG AAT GAA GCC CAC
GAPDH	Forward: GCC AAG GTC ATC CAT GAC AAC
	Reverse: GTC CAC CAC CCT GTT GCT GTA

Immunofluorescent Staining

iPS cells are fixed with 4% paraformaldehyde in phosphate bufered saline (PBS) for 15 min at 4 °C. After washing with PBS, the cells are treated with 0.2% Tween-20 in PBS for 5 min at room temperature, blocked with 5% goat serum in PBS for 30 min at room temperature and then stained with OCT-4, SSEA-4, TRA-1-60, TRA-81 primary antibodies. . Primary monoclonal antibodies include SSEA-4, TRA-1-60, TRA-1-81, OCT-4, SOX2 (Milipore, USA, 1:500 dilution, respectively).

Secondary antibodies used are Alexa568-conjugated goat anti-mouse IgM (1:500), Alexa568- conjugated goat anti-mouse IgG (1:500). Cell nuclei are counter stained with 2 μg/ml DAPI for 30 min (Invitrogen, USA).

RT-PCR

Total RNA of cultured cells is extracted with Trizol reagent (Invitrogen, USA) and purified with RNeasy™ Kit to remove genomic DNA contamination following the manufacturer's protocol. 500 ng of total RNA is used for reverse transcription reaction with Oligo dT and dNTP using the Superscript III kit (Invitrogen, USA). RT-PCR is performed with Platinum PCR master mix (Invitrogen, USA) and PCR products are subjected to agarose gel electrophoresis for detecting of PCR products. All primers used are shown in Table 1.

Embryoid Body- Mediated Differentiation

Human iPS cells are harvested with collagenase IV. The cells are cultured in a floating suspension containing DMEM/F12 with 20% knockout serum replacement (Invitrogen, USA), 2mM L-glutamine (Invitrogen, USA), 1×10^{-4} M nonessential amino acids, 0.1 mM 2-mercaptoethanoel (Invitrogen, USA) and 0.5% penicillin and streptomycin. After 8 days culture, the EBs formed are transferred to gelatin-coated plate and cultured in the same medium for another 7 days.

Teratoma Formation

iPS cells are harvested using collagenase IV treatment from a 6cm petri dish and are centrifuged at 200g for 5 minutes. The iPS cell pellet is re-suspended in DMEM/F12. The iPS cells are injected subcutaneously into dorsal flank of 6-8 weeks old SCID mice (Jackson lab, USA).

Eight weeks after injection, SCID mice are sacrificed, tumors are dissected and fixed with 10% formalin overnight. The tumor is embedded in paraffin and paraffin sections are stained with hematoxylin and eosin.

Differentiation of Induced Pluripotent Stem Cells into Functional Cardiomyocytes

iPS cells in tissue culture plate are treated with 2 unit/ml of Dispase (Roche) for 15 minutes at 37°C to detach the iPS cell colonies. The detached iPS cell colonies are collected, and suspended in ultra-low 6-well culture plate (Corning life Sciences) with iPS cell culture media without bGF to form embryoid bodies (EBs). The next day the culture media is changed to EB20 (DMEM containing 20%FBS, 1 mM non-essential amino acids, 1 mM l-glutamine, 0.1 mM β-mercaptoethanol, 100 unit/ml penicillin, 100 μg/ml streptomycin), after culturing in supsension for 7 days, the EBs are plated onto 100-mm petri dishes coated with 0.1% gelatin solution in EB20 media. 7-10 days after EBs plated onto gelatin coated dishes, the culture medium is chagned to EB2 (DMEM containing 2%FBS, 1 mM non-essential amino acids, 1 mM l-glutamine, 0.1 mM β-mercaptoethanol, 100 unit ml−1 penicillin, 100 μg ml−1 streptomycin). The EBs will begin to contract spontaneously 2-3 weeks after plating. 2 weeks after EBs begin to beat, the beating EBs are isolated manually using 22G needle, placed into 15ml tubes, washed once with PBS, incubated in 0.25% trypsin-0.02% EDTA (Invitrogen) for 4 minutes at 37°C. Single cells are dissociated form EBs, using a P200 filter tip to pipette up and down. The single cells are plated onto gelatin-coated coverslips in 24-well plate.

Characterization of Cardiomyocytes

Immunofluorescent Staining

The identity and structural integrity of CMs are determined by demonstrating the presence of cardiac contractile protein α-actinin and β-myosin heavy chain (β-MHC). The beating cardiomyocytes are isolated and fixed with 4% paraformaldehyde in PBS for 15 min at 4 °C. After washing with PBS, the cells and EBs are treated with 0.2% Tween-20 in PBS for 5 min at room temperature and incubated with anti α-actinin (Sigma, USA), anti β-myosin heavy chain and anti titin (Millipore, USA) antibodies. Secondary antibodies used are Alexa568-conjugated goat anti-mouse IgM (1:500) and Alexa568- conjugated goat anti-mouse IgG (1:500). Cell nuclei are counter stained with 2 μg/ml DAPI for 5 min (Invitrogen, USA). The stained cell cultures are subjected to Laser Confocal microscopy (LSM710, Carl Zeiss) for taking pictures.

RT-PCR

Total RNA of beating clusters are extracted with Trizol reagent (Invitrogen, USA) and purified with RNeasy™ Kit to remove genomic DNA contamination following the manufacturer's protocol. 500ng of total RNA is used for reverse transcription reaction with Oligo dT and dNTP using the Superscript III kit (Invitrogen, USA). RT-PCR is performed with Platinum SYBR Green qPCR Supermix (Roche Applied Science, Switzerland) and analyzed with the 7300 real-time PCR system (Applied Biosystems, USA). All primers used are shown in Table 1.

Single Cell Patch Clamp

For the patch clamp studies, spontaneously beating small cell clusters are isolated and plated on top of gelatin-coated glass coverslips. The patch pipette solution contains (mM): 120 KCl, 1 MgCl2, 3 Mg-ATP, 10 HEPES, 10 EGTA (pH 7.3). The bath solution contains (mM): 140 NaCl, 5.4 KCl, 1.8 CaCl2, 1 MgCl2, 10 HEPES, 10 glucose (pH 7.4) (all materials from Sigma-Aldrich). Action potentials are recorded using the current clamp mode. Axopatch 200B, Digidata 1322, and pClamp10 (Molecular Devices, Sunnyvale, CA, USA) are used for data amplification, acquisition, and analysis.

Laser Confocal Ca^{2+} Imaging for Determining Ca^{2+} Transients

Calcium imaging using confocal fluorescent microscopy is conducted on CMs that are pre-loaded with 6 μg/mL Fluo-4 AM (Molecular Probes) for 15 min at 37°C and maintained in normal Tyrode solution. Ca^{2+} transients are recorded by a LSM-710 laser scanning confocal microscope (Carl Zeiss, Germany) with a × 40 oil immersion objective, numeric aperture = 1.3. Fluo-4 is excited at 488 nm using a 25 mW argon laser (with intensity attenuated to 1%). Fluorescence emission is measured at > 505 nm. Images are acquired in the line (X-T)-scan mode with 512 pixels per line at a rate of 3 ms per scan. The scan line is oriented along the longitudinal axis of the cell, at pixel intervals of 0.15 μm. The axial resolution is set at 1.5 μm. Ca^{2+} transients are measured with the confocal microscope operating in the frame (X-Y) imaging mode. Ca2+ images are analyzed using a computer program written in IDL 5.4 software.

Anticipated Results

Skin Fibroblasts Culture and Derivation of iPS Cells

Human 6-mm skin punch biopsies are minced and dissociated dermal fibroblasts are plated to expand dermal fibroblasts (Figure 3A). Fibroblasts are infected with retroviral viruses containing 4 transcription factors genes (Oct4, Sox2, C-Myc and Klf4). Infected cells are incubated in human embryonic stem (hES) cell medium until iPS cell colonies start to appear (Figure 3B). iPS cell colonies are picked and expanded for further use (Figures 3C and 3D).

Characterization of iPS Cells

Induced pluripotent stem (iPS) cells exhibit the following characteristics similar to those of hES cells: (1) iPS cells from human dermal cells express pluripotent genes (Figure 4A); Endogenous genes are reactivated, while transgenes in some clones are silenced (Figure 4B); (2) iPS cells show staining of the pluropotent markers of hES cells (Figure 2C); (3) Formation of EBs by floating culture of iPS cells (Figure 4D); and (4) iPS cells form tetratoma after injection into SCID mice for 8 weeks (Figure 4E).

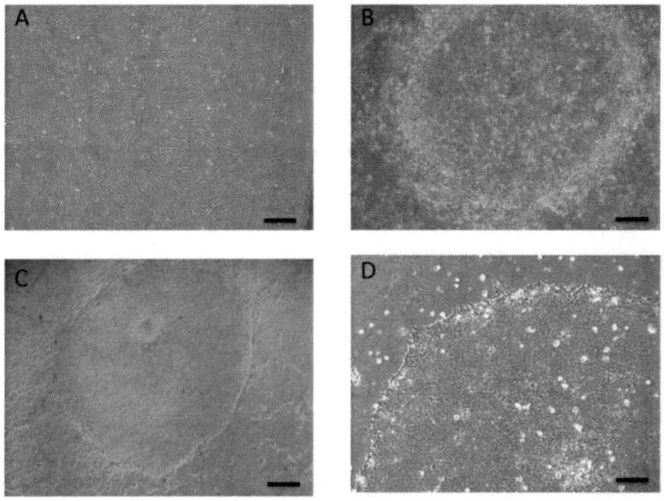

Figure 3. Human dermal fibroblasts and induced pluripotent stem cells. (A) The primary culture of human dermal fibroblasts. (B) Dermal fibroblasts infected with retroviruses (Oct4, Sox2, C-Myc and Klf4) were plated onto MEF feeder. Embronic stem (ES) cell-like iPS cell colonies appear 3-4 weeks after plating. (C) The iPS cell colonies were manually picked and subcultured onto MEF feeder. (D) iPS cells at passage 4 were grown on Matrigel in mTeSR1 medium. Scale bars: 3A, 3B and 3C, 20μm; 3D: 50μm.

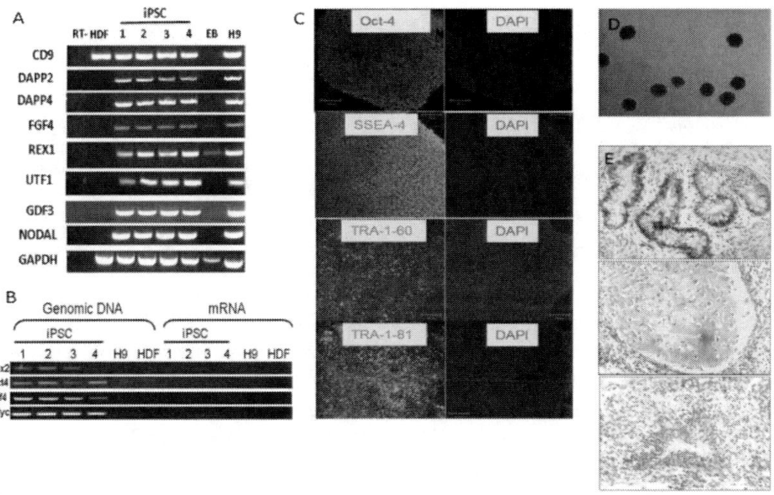

Figure 4. Characterization of iPS cells. (A) RT-PCR analysis of pluripotent genes expression in iPS cells. The high level expression of pluripotent genes were observed in iPS cell lines (1-2), which is simlar to those of human ES cell (H7), RT- was used as negative control, while dermal fibroblast was used as controls. (B) PCR analysis of the expression of retroviral transgenes in iPSCs. Genomic DNA and mRNA were extracted from iPSC clones 1 to 4 . Retroviral transgenes of Sox2, Oct-4, Klf4 and C-Myc were only detected in genomic DNA of iPSCs. (C) Immunofluorescent staining of iPSC markers. Positive staining of Oct-4 (red) was observed in cell nuclei while positive staining of SSEA-4, TRA-1-60, and TRA-1-80 (red) were observed in cell membrane. Cell nuclei were counter stained with DAPI (blue). (D) iPS cells were cultured in floating culture and embryoid bodies were formed 1 week later. (E) Teratoma forming assay of iPS cells. iPS cells were injected into kidney capsules of 6-8 weeks old SCID mice. Tumors were harvested 10 weeks later and subjected to H & E staining. Teratomas included the tissues of 3 primative germ layers, such as glandular structure for endoderm, cartilage for mesoderm, and neuroepithelium for ectoderm.

Differentiation of Cardiomyocytes from iPS Cells

Spontaneously contracting EBs emerge 2-3 weeks after plating EBs onto 0.1% gelatin-coated dishes. Immunofluorescence microscopy with antibodies specific for myofilament proteins determines whether cells exhibit organized sarcomeres proteins (β-myosin heavy chain, α-actinin and titin) that are specific for CMs (Figure 5A). Functional analysis of CMs can be performed by cellular electrophysiology measurements to determine whether pacemaker-, ventricular, and atrial-like action potentials are present (Figure 5B). Detection of Ca^{2+} transients specific for CMs provides assessment of the functional integrity of the differentiating CMs (Figure 5C).

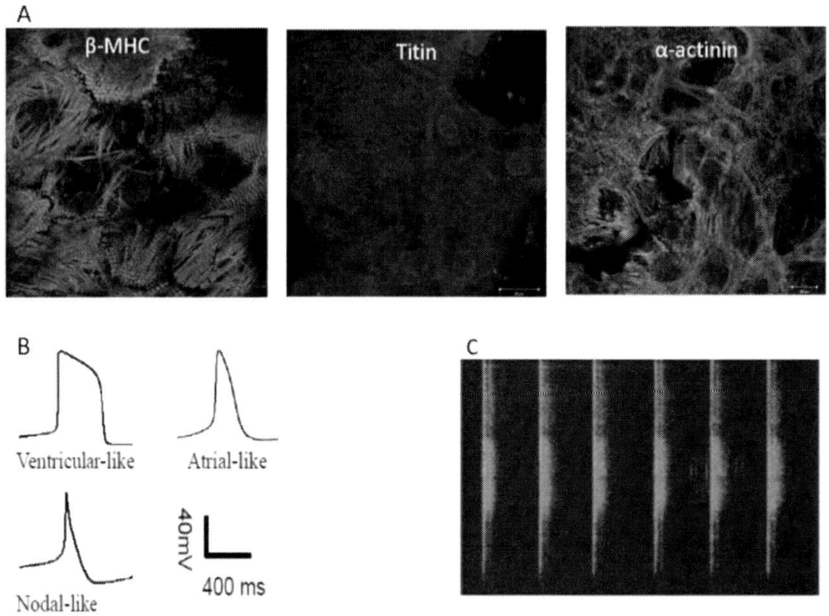

Figure 5. Characterization of cardiomyocytes derived from iPS cells. (A) Fluroscent images of cardiac-specific marker in cardiomyocytes derived from iPSCs using following immunofluorescence staining. The expression of β-myosin heavy chain (β-MHC, green), titin (red), and α-actinin (green) were observed in cytoplasm. The cell nuclei were counter stained with DAPI (blue). (B) Action potential recording in iPS cell-derived caridomyocytes. Atrial-, nodal- and ventricular- action potentials were recorded in iPS cell derived cardiomyocytes. (C) Calcium transients recorded from iPSC derived cardiomyocytes.

Conclusion

Robust protocols of generating iPS cells and differentiating iPS cells to CMs are essential to advance biomedical research and clinical applications of iPSCs. For example, the research using cardiac disease-specific and patient-specific iPS cell lines will be greatly facilitated by the availability of cardiac differentiation protocols that are applied in iPSC lines and do not require cell line-specific optimization. Specifically, iPSC-derived CMs can be used for transplantation for replacement of CMs lost from natural cardiac disease processes or pathology. For the cardiomyocytes derived from iPS cells using current protocol, the cardiomyocytes are heterogeneous even within a single beating cluster; and these cardiomyocytes are not fully differentiated even after differentation for a few weeks. Therefore, future studies are needed to develop the protocol for generating homogenous populations of specific CM types and enhancing the maturation of these cells before these clinical applications can be fully realised.

References

Anokye-Danso, F., Trivedi, C.M., Juhr, D., Gupta, M., Cui, Z., Tian, Y., Zhang, Y., Yang, W., Gruber, P.J., Epstein, J.A., *et al.* (2011). Highly efficient miRNA-mediated reprogramming of mouse and human somatic cells to pluripotency. *Cell Stem Cell 8*, 376-388.

Chang, C.W., Lai, Y.S., Pawlik, K.M., Liu, K., Sun, C.W., Li, C., Schoeb, T.R., and Townes, T.M. (2009). Polycistronic lentiviral vector for "hit and run" reprogramming of adult skin fibroblasts to induced pluripotent stem cells. *Stem Cells 27*, 1042-1049.

Fusaki, N., Ban, H., Nishiyama, A., Saeki, K., and Hasegawa, M. (2009). Efficient induction of transgene-free human pluripotent stem cells using a vector based on Sendai virus, an RNA virus that does not integrate into the host genome. *Proc Jpn Acad Ser B Phys Biol Sci 85*, 348-362.

Inoue, H., and Yamanaka, S. (2011). The use of induced pluripotent stem cells in drug development. *Clin Pharmacol Ther 89*, 655-661.

Itzhaki, I., Maizels, L., Huber, I., Zwi-Dantsis, L., Caspi, O., Winterstern, A., Feldman, O., Gepstein, A., Arbel, G., Hammerman, H., *et al.* (2011). Modelling the long QT syndrome with induced pluripotent stem cells. *Nature 471*, 225-229.

Kim, D., Kim, C.H., Moon, J.I., Chung, Y.G., Chang, M.Y., Han, B.S., Ko, S., Yang, E., Cha, K.Y., Lanza, R., *et al.* (2009). Generation of human induced pluripotent stem cells by direct delivery of reprogramming proteins. *Cell Stem Cell 4*, 472-476.

Ma, D., Wei, H., Lu, J., Ho, S., Zhang, G., Sun, X., Oh, Y., Tan, S.H., Ng, M.L., Shim, W., *et al.* (2012). Generation of patient-specific induced pluripotent stem cell-derived cardiomyocytes as a cellular model of arrhythmogenic right ventricular cardiomyopathy. *Eur Heart J.*

Miyoshi, N., Ishii, H., Nagano, H., Haraguchi, N., Dewi, D.L., Kano, Y., Nishikawa, S., Tanemura, M., Mimori, K., Tanaka, F., *et al.* (2011). Reprogramming of mouse and human cells to pluripotency using mature microRNAs. *Cell Stem Cell 8*, 633-638.

Moretti, A., Bellin, M., Welling, A., Jung, C.B., Lam, J.T., Bott-Flugel, L., Dorn, T., Goedel, A., Hohnke, C., Hofmann, F., *et al.* (2010). Patient-specific induced pluripotent stem-cell models for long-QT syndrome. *N Engl J Med 363*, 1397-1409.

Oh, Y., Wei, H., Ma, D., Sun, X., and Liew, R. (2012). Clinical applications of patient-specific induced pluripotent stem cells in cardiovascular medicine. *Heart 98*, 443-449.

Okita, K., Ichisaka, T., and Yamanaka, S. (2007). Generation of germline-competent induced pluripotent stem cells. *Nature 448*, 313-317.

Okita, K., Nakagawa, M., Hyenjong, H., Ichisaka, T., and Yamanaka, S. (2008). Generation of mouse induced pluripotent stem cells without viral vectors. *Science 322*, 949-953.

Okita, K., and Yamanaka, S. (2010). Induction of pluripotency by defined factors. *Exp Cell Res 316*, 2565-2570.

Si-Tayeb, K., Noto, F.K., Sepac, A., Sedlic, F., Bosnjak, Z.J., Lough, J.W., and Duncan, S.A. (2010). Generation of human induced pluripotent stem cells by simple transient transfection of plasmid DNA encoding reprogramming factors. *BMC Dev Biol 10*, 81.

Somers, A., Jean, J.C., Sommer, C.A., Omari, A., Ford, C.C., Mills, J.A., Ying, L., Sommer, A.G., Jean, J.M., Smith, B.W., *et al.* (2010). Generation of transgene-free lung disease-specific human induced pluripotent stem cells using a single excisable lentiviral stem cell cassette. *Stem Cells 28*, 1728-1740.

Stadtfeld, M., Nagaya, M., Utikal, J., Weir, G., and Hochedlinger, K. (2008). Induced pluripotent stem cells generated without viral integration. *Science 322*, 945-949.

Takahashi, K., Tanabe, K., Ohnuki, M., Narita, M., Ichisaka, T., Tomoda, K., and Yamanaka, S. (2007). Induction of pluripotent stem cells from adult human fibroblasts by defined factors. *Cell 131*, 861-872.

Warren, L., Manos, P.D., Ahfeldt, T., Loh, Y.H., Li, H., Lau, F., Ebina, W., Mandal, P.K., Smith, Z.D., Meissner, A., *et al.* (2010). Highly efficient reprogramming to pluripotency and directed differentiation of human cells with synthetic modified mRNA. *Cell Stem Cell 7*, 618-630.

Woltjen, K., Michael, I.P., Mohseni, P., Desai, R., Mileikovsky, M., Hamalainen, R., Cowling, R., Wang, W., Liu, P., Gertsenstein, M., *et al.* (2009). piggyBac transposition reprograms fibroblasts to induced pluripotent stem cells. *Nature 458*, 766-770.

Yazawa, M., Hsueh, B., Jia, X., Pasca, A.M., Bernstein, J.A., Hallmayer, J., and Dolmetsch, R.E. (2011). Using induced pluripotent stem cells to investigate cardiac phenotypes in Timothy syndrome. *Nature 471*, 230-234.

Yu, J., Hu, K., Smuga-Otto, K., Tian, S., Stewart, R., Slukvin, II, and Thomson, J.A. (2009). Human induced pluripotent stem cells free of vector and transgene sequences. *Science 324*, 797-801.

Yu, J., Vodyanik, M.A., Smuga-Otto, K., Antosiewicz-Bourget, J., Frane, J.L., Tian, S., Nie, J., Jonsdottir, G.A., Ruotti, V., Stewart, R., *et al.* (2007). Induced pluripotent stem cell lines derived from human somatic cells. *Science 318*, 1917-1920.

Zhou, H., Wu, S., Joo, J.Y., Zhu, S., Han, D.W., Lin, T., Trauger, S., Bien, G., Yao, S., Zhu, Y., *et al.* (2009). Generation of induced pluripotent stem cells using recombinant proteins. *Cell Stem Cell 4*, 381-384.

Zhou, W., and Freed, C.R. (2009). Adenoviral gene delivery can reprogram human fibroblasts to induced pluripotent stem cells. *Stem Cells 27*, 2667-2674.

Index

C

D

E

F

G

H

I

J

K

L

M

N

O

P

Q

R

S

T

U

V

W

X

Y

Z